CRAFT OF THE DYER

Karen Leigh Casselman

Craft of the Dyer: Colour from Plants and Lichens of the Northeast

UNIVERSITY OF TORONTO PRESS Toronto Buffalo London

Toronto Buffalo London
Printed in Canada

Canadian Cataloguing in Publication Data

Casselman, Karen Leigh, 1942–
Craft of the dyer
Bibliography: p.
Includes index.
ISBN 0-8020-2362-2
1. Dyes and dyeing, Domestic. 2. Dye plants – Canada, Eastern.* 3. Dye plants – Northeastern States. I. Title.
TT854.3.C37 667'.26 C80-094168-3

Publication of this book has been assisted by the Canada Council and the Ontario Arts Council under their programs of block grants.

TO MY FAMILY

Contents

Preface

Although an understanding of nature is not a prerequisite for the would-be dyer, our physical surroundings form an integral part of the dyeing process. Dyers develop an intuitive interest in and concern for the environment. The dyer learns to know the natural world, and so gains insights into the contemporary application of traditional dyeing techniques to modern craft forms.

The dyer begins to sense the seasons, interpret nature's whims, and respond to the subtleties of natural change and renewal. Powers of observation seem enhanced. The senses are more attuned to colours, smells, and textures. Form appears as design. One may notice, for the first time, that lichens are not mosses. Wild radish blooms on into November in many parts of northeast North America (east of Manitoba and Minnesota and as far south as Virginia), making it possible to use this common weed for dyeing long after garden flowers have succumbed to frost. Nuts and bark can be collected for dyeing in the winter, and the kitchen yields abundant dyestuffs when the weather prevents outdoor collecting. A plant that arouses the curiosity of one dyer will be bypassed by another, the very diversity which is so typical of plant dyeing. It is truly an interpretive skill. The only tradition that has remained unchanged is the love of doing it.

Acknowledgements

I am grateful for the support and assistance of the following people: Marie Aiken, Joan Auld, Leona Banks, Barbarie Bethune, Greg Cook, Marian deWitt, Pauline Diadick, Fred Diadick, Charles Doucet, Carol Duffus, Geri Gaskin, Lesleigh Grice, Mary Hill, Ann Hillis, Valerie Kenny, Brenda Large, Orland Larson, Dawn MacNutt, Alicia Marr, Patricia McClelland, John Eaton McClelland, Debra Pollock, Sheila Simpson, Ken Sonnenburg, Mary Sparling, Cynthia Tanner, Dwight Tanner, and Alex Wilson.

Edgar Bennett, Marilyn MacDonald, and Ron Shuebrook supported other professional activities which allowed me to carry on with my research and manuscript preparation.

Special thanks are owed to the Canada Council Explorations Program, to all my students, and to Maritime Colour Lab, Dartmouth NS.

KC

CRAFT OF THE DYER

Introduction

There are numerous myths and misconceptions associated with dyeing, which may discourage the beginning dyer. Should an enamel or aluminum pot be used? Is it necessary to premordant fibres to be dyed? How much of a mordant is required? Questions like these worry novices and lessen their enthusiasm for dyeing. None of the answers to these questions is complicated, but to obtain them one must often read many books and articles on plant dyeing. Reference material written by authors living and working in other countries can be unintentionally misleading. For example, goldenrod, which is a common weed in Canada, is cultivated as a garden flower in parts of England. Although foreign texts are invaluable research aids to the serious dyer, beginning dyers should look to their own geographical region for instructional material appropriate to their plant dyeing.

In order to facilitate the exchange of information among dyers, Latin names are given for all dye sources listed in this book. Too often a dyer meets someone from another country who shares an interest in dyeing, but is unable to communicate. '*Pinus*' is pine, and '*Rhus*' is sumac, no matter where a dyer lives.

The information presented in this book has been collected, researched, and tested since 1974. Nothing rare is suggested as a potential source of dye. Imported dyestuffs are not dealt with because the priority now appears to be for dyers to become more aware of plant life available immediately around them. One exception is the imported blue-yielding indigo. A recipe is given for its use because few other sources of blue are as satisfactory. Without indigo, the dyer's colour vocabulary would be severely restricted. (See also copper penny blue, p 127, and woad, p 220.)

The amount of yarn given in each recipe (one pound, 453 g of fibre) is based on the belief that the time-consuming process of dyeing warrants using enough yarn to be sufficient for an entire craft project, however small that article may be. Six skeins of 2-ply wool yarn, weighing four ounces (114 g) each would enable the dyer to weave a throw, crochet a shawl, or knit a sweater. While it is important to keep records and a swatch book containing dyed yarn samples, these aspects of dyeing should not be more important to the dyer than collecting dyestuffs and using dyed fibre. The investigation of plant dyeing is enhanced by the dyer's sympathetic use of the colours produced, whether it be for macramé, embroidery, crewel, knitting, crochet, or weaving.

Just as one cannot predict the exact shade maple leaves will turn in October, neither can anyone working with plant dyes guarantee the colour that will result from use of a specific plant. There are too many variables; indeed, it is these subtle influences that make dyeing exciting. A dyer should aim to create new shades, rather than simply to duplicate others' results. In this way even beginning dyers can add to existing information on plant dyeing, thereby making a contribution of benefit to all.

Approach dyeing as you would the making of an omelette – the delight is in knowing that the dish never will be quite the same. The dyer's recipe book is as private a domain as the cook's kitchen. Respect the information it offers, alter it if need be, and remember to share what you have learned. No brilliant colour was ever known to fade because it gave joy to yet another craftsperson.

Handspun yarn dyed with ONION, using a variety of mordants. Clockwise, starting at upper right: (gold) no mordant, medium strength bath; (rust) chrome; (grey-green) iron; (brown) chrome and blue vitriol; (lemon-yellow) no mordant; (beige) blue vitriol, weak bath; (dark grey-brown) chrome and iron, strong bath; (orange) tin; (light grey) iron exhaust; (light rust-brown) chrome exhaust

Icelandic Lopi wool, dyed in UMBILICARIA baths, without heat. From the top: (rose) second dip in vinegar bath; (grey-orchid) third dip in magenta bath; (light rose-beige) third vinegar exhaust dip; (magenta) no mordant

Icelandic Lopi wool, dyed with weeds. From the left: (dark green) BURDOCK with iron; (green) MULLEIN with blue vitriol; (gold) MULLEIN with alum; (rust) WILD ASTER with chrome

Cloth woven from twelve shades of bouclé yarn dyed with UMBILICARIA. The bright magenta was processed in a regular heated bath, with no mordant; the strongest red in a heated bath with vinegar; the lighter colours are from exhaust dips

Herringbone twill woven from wool/acrylic/mohair blend yarns dyed with UMBILICARIA (red and pink) and LOGWOOD (dark grey), with heat

top: ONION, no mordant; WHITE ASTER, chrome; QUEEN ANNE'S LACE, alum
centre: TANSY, chrome and iron; SWEET FERN, blue vitriol; BURDOCK, blue vitriol
bottom: DAY LILY, chrome; ONION, chrome; MULLEIN, blue vitriol and iron

top: BURDOCK, blue vitriol; UMBILICARIA on beige yarn, no mordant: SWEET FERN, tin
centre: UMBILICARIA on grey yarn, no mordant; LAUREL, iron; UMBILICARIA, third exhaust, no mordant
bottom: ONION, alum; COCHINEAL, first exhaust; POPLAR LEAVES, iron

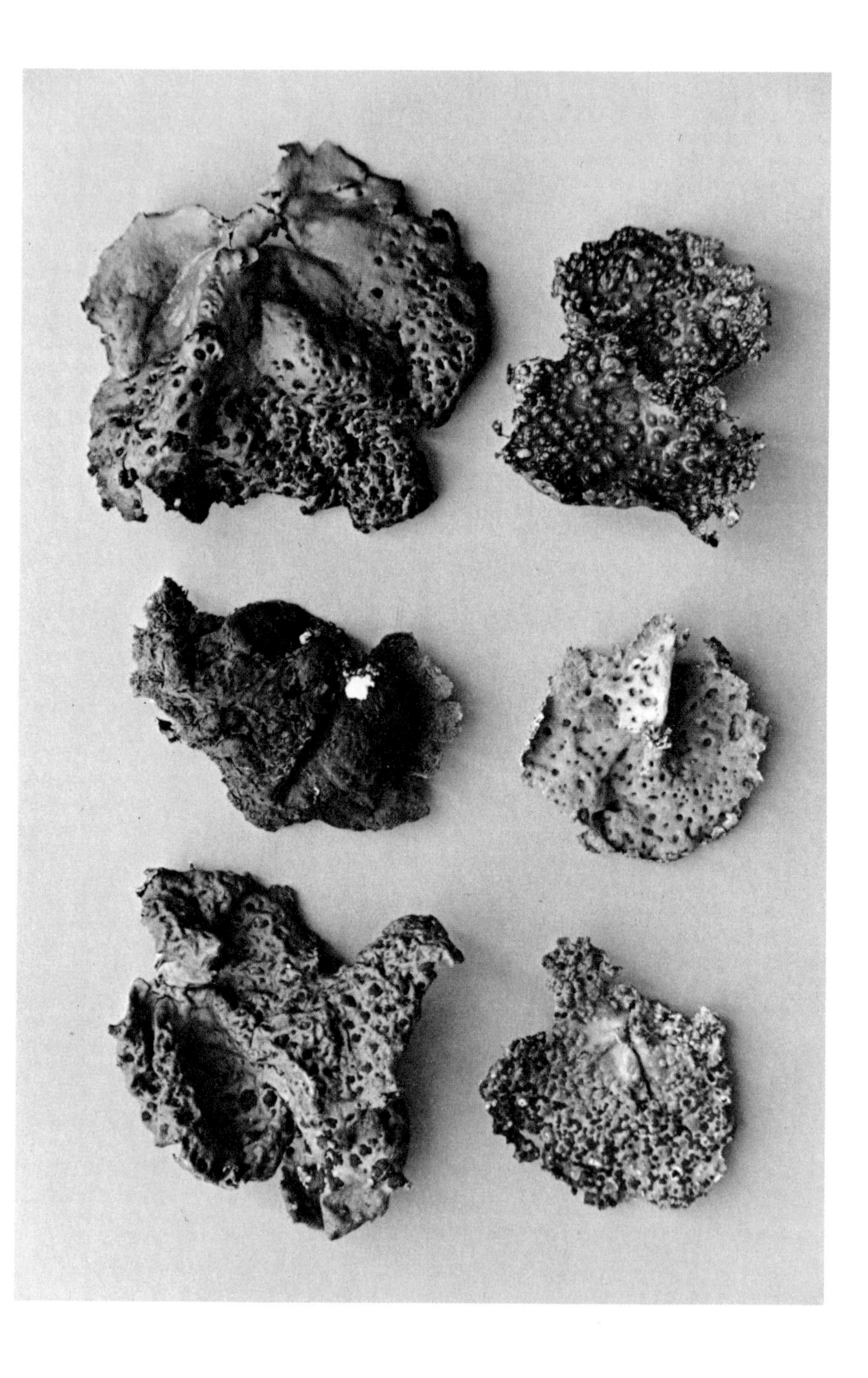

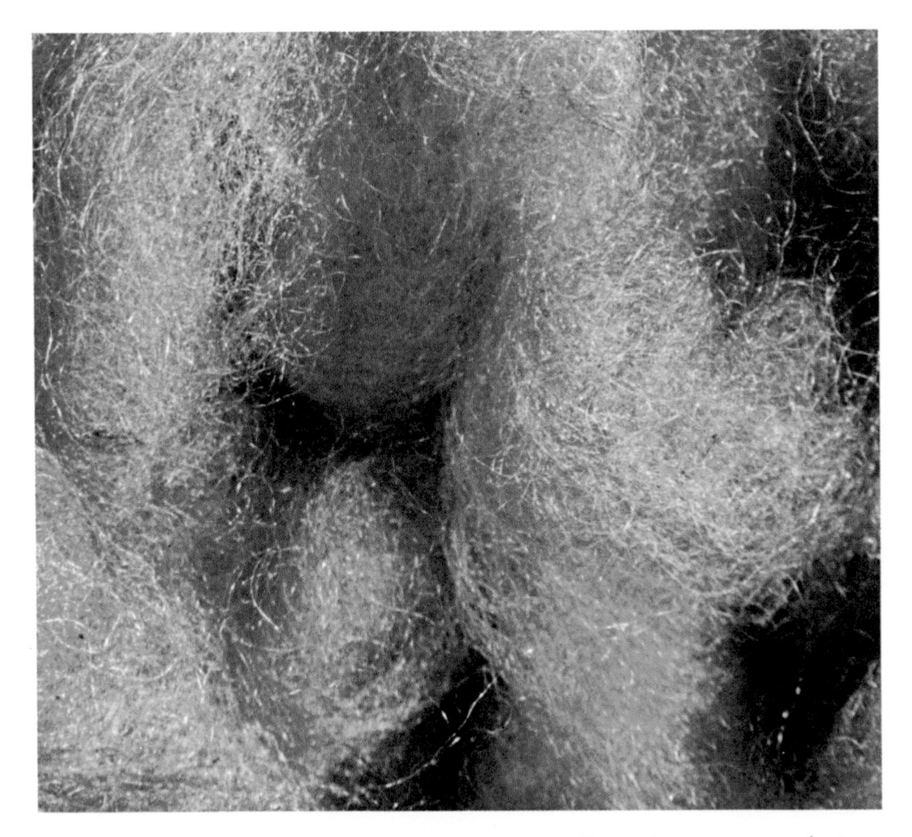

Carded Leicester fleece, ready for spinning. Dyestuffs used were LAUREL (yellows) and UMBILICARIA (pinks).

OPPOSITE Two species of UMBILICARIA
left: *U. mammulata*
top right side up, wet thallus colour; *centre* turned over, showing black velvety lower cortex; *bottom* right side up, dry thallus colour
right: *U. deusta*
top right side up, wet thallus colour; *centre* turned over, showing fawn lower cortex; *bottom* right side up, dry thallus colour

NB The dry thallus colour of this sample of *U. mammulata* is brown, but some may be much darker (brown-black). The wet thallus colour is often olive-green. The colours of *U. deusta* shown here are very true of most specimens; there is not much variance with this species.

Handspun yarn. From the top: ONION, no mordant; HEAL-ALL, iron; BIRCH BARK, no mordant

Handspun yarn. The brown yarn at the top was dyed with HEAL-ALL. The blue yarns were all dyed with INDIGO: *top*, single-ply, 1 dip in indigo; *centre*, 2-ply, 4 dips in indigo; *bottom*, 2-ply, exhaust dip

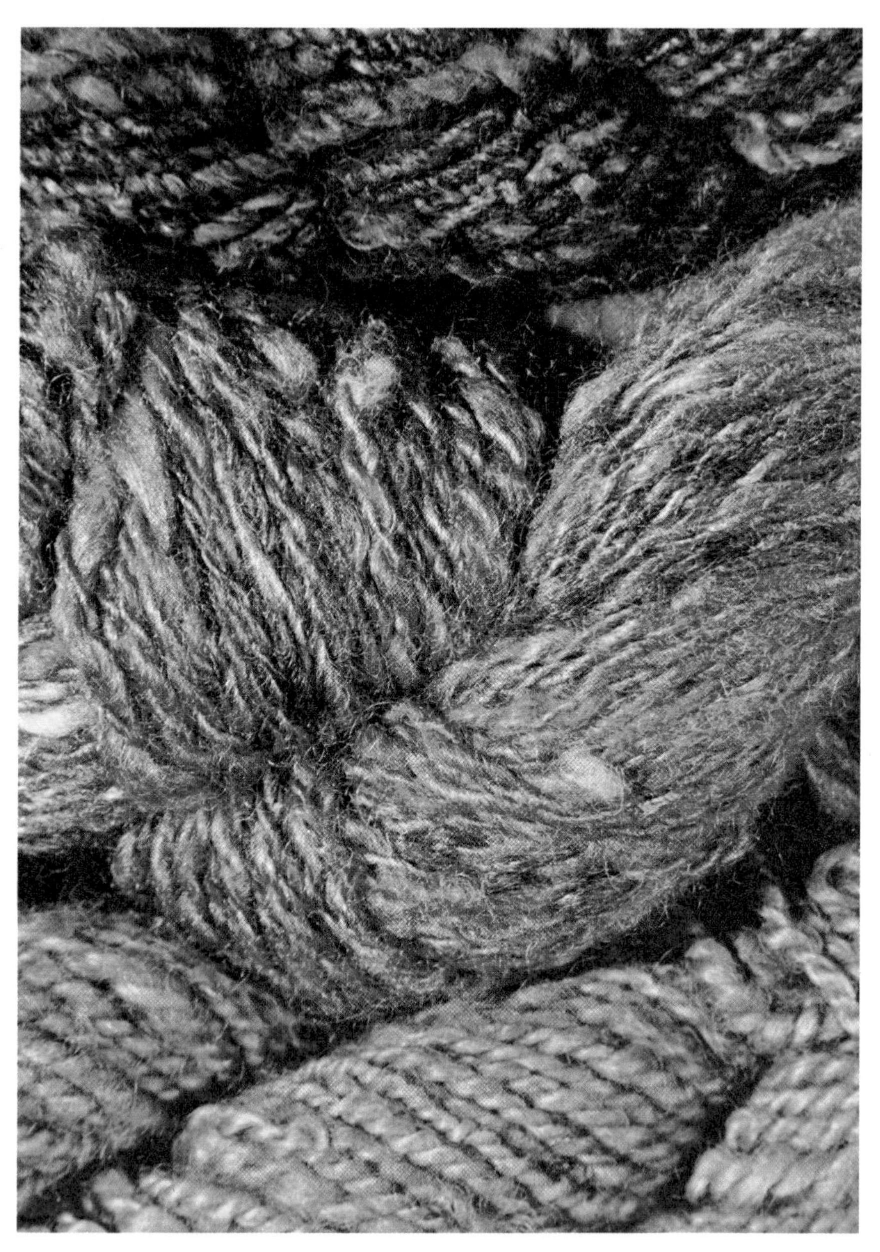

Handspun yarn. The two blues were dyed with INDIGO, after spinning. The multi-coloured skein in the middle was dyed in the fleece.

I

Why bother?

It is often suggested that making dyes from plants is a highly specialized field which belongs more in the realm of the textile purist than that of the contemporary fibre artist and craftsperson. Likened to the earnest gourmet who grinds her own flour for bread, the dyer has been viewed as someone who experiments with dyes but rarely enjoys them. Nothing could be further from the truth. Increasingly, modern craftspersons are turning to natural dyes as a logical extension of their aesthetic concerns in weaving, macramé, spinning, hooking, knitting, crocheting, and stitchery. Quilters and home-sewers are experimenting with plant-dyed fibres, often with unexpected success. Furthermore, it is encouraging to see that plant dyes are now as often made by schoolchildren in their classrooms as by folklorists and devotees bent on relearning earlier skills.

Process is the key to making dyes. The techniques used to extract pigment from plants and to make the resulting colour are not unlike dressing the loom or wedging clay prior to throwing on the wheel. Each step is important. These separate steps then blend into a loose schematic structure that makes results significant. Sometimes the process is unstructured, as is often the case in the elementary school classroom. On the other hand, the sophisticated and controlled dyeing experiments of the advanced student of any of the fibre arts require careful attention to detail, a thorough knowledge of procedures, and somewhat specialized equipment. What matters is not which techniques are used, for there are many, but that the results satisfy the dyer. Successful dyeing means that the total experience was worthwhile.

Making dyes from plants can expand the whole concept of fibre and its contemporary craft application. On a personal level, plant dyeing greatly affects one's personal response to colour. Even the most inexperienced dyers acknowledge that they become less satisfied with ready-dyed yarns in shops after they have learned to produce one or two shades of their own. No two dyebaths are ever exactly the same. This means that novice as well as advanced dyers can rightly claim colours they develop as 'original.' It is this built-in authenticity which appeals to the discriminating craftsperson – the fact that no one else can duplicate a piece of their work made from plant-dyed fibres. Such yarns add a touch of distinction and suggest to the viewer that the designer/maker fully developed his creative concept from start to finish, although the design itself may be random or abstract. But the dyer need not worry that using commercially-dyed fibres is heretical. Indeed, working with plant dyes helps the craftsperson to be more selective in making choices which influence his or her work. Many yarns available today at specialty and weaving shops are extremely handsome and a delight to use. The perceptive dyer will soon discover what yarn belongs where, no matter what its origin.

Plant-dyed fibres and the dyeing process itself are stimulants to creativity. An entire concept for a new piece of work may well be formulated before the freshly-dyed yarn is dry. The nuances of colour lead the dyer to consider new combinations and unexplored juxtapositions, for plant-dyed shades rarely clash or jar the eye. Their natural harmony is a special kind of perfection that appeals to those who have learned to trust their senses.

Why bother to make dyes from plants? It is more convenient to purchase ready-dyed fibres, and easier still to duplicate the colour schemes featured in textile books, but using others' ideas does not help the serious craftsperson to be individual, any more than it benefits the adventuresome knitter always to rely on printed patterns.

Making dyes requires time, energy, and patience, but the rewards are sufficiently gratifying that those who try to use plant-dyed fibres will have a strong sense of originality. 'Doing things from scratch' used to be a source of pride when problem-solving was part of each day's activities. Whatever was needed, one made. Whatever did not work was replaced by an artifact that did. Necessity gave rise to many ingenious work methods. Because it often encourages spinning, making dyes takes one back to that simpler approach of making something 'from scratch.'

Woven cloth of plant-dyed fibres has an unmistakeable heathery quality, which reminds some people of fog. The shades are subtle and harmonize with each other in such a way as to give the fabric great depth of colour. 'Sweet, beautiful colours' is the description of their own dyes by some Eskimo women who worked at Spence Bay, NWT, with dyer Judy McGrath. (See Judy McGrath, 'The Dye Workshop.') Each plant provides an amazing diversity of shades: onion skin-dyed fibre may be beige, soft yellow, lemon yellow, bright gold, old gold, orange, bronze, brass, khaki, avocado, tan, warm brown, rust, burnt orange, or russet. That is just the beginning. To use even three or four of the above shades in a single craft project will result in the finished article having an uncomparably rich look. You will never again fall into the trap of thinking, 'Yellow goes with brown, beige goes with white.'

TERMINOLOGY

The terminology used in this book conforms to the standard North American usage of words and phrases associated with plant dyeing. Making dyes from plants is herein called 'plant dyeing,' a term which means the same thing as 'natural dyeing,' 'organic dyeing,' and 'vegetable dyeing.' 'Plant dyeing' is considered the most appropriate term, as the use of chemicals makes the term 'natural' a misnomer. Most dyestuffs are vegetable matter (flowers, weeds, barks, lichens, and so on). Among the few animal dyestuffs are blood and the red-yielding, imported cochineal. Mineral dyestuffs include the traditional iron buff and copper penny blue (p 127).

acid organic and inorganic compounds which are often used in dyeing (vinegar, tannic acid, cream of tartar) to alter the pH value of the dyebath; acids neutralize alkaline (base) substances and change litmus from blue to red. When added to the dyebath, acids soften the water and reduce the possible harmful effects of alkalis such as lye and ammonia.

adjective dyestuff which requires a mordant to affix the colour and make it fast

alkali compounds which are used in dyeing to neutralize acids and alter dyebaths from red to blue (particularly *Umbilicaria* baths). Ammonia, alum, and lye are alkaline, and the following substances also function as such in the dyepot: wood ash; sal soda (washing soda); baking soda; detergent and lye-based soaps.

base alkali

bleach sodium, or calcium hypochlorite, abbreviated to CL (indicating the active ingredient, chlorine). To bleach a fibre is to remove its natural colour. A bleached wool yarn appears 'snow white' compared with the ivory or greyish-white of a similar but 'unbleached' fibre.

boil Liquid at a temperature of 98–100°C or 212°F. For convenience, the celsius equivalent to 212°F is herein given as 100°C. (See simmer.)

cook out to boil a dyestuff in water until the colour is extracted, the resulting liquid constituting the 'dyebath'

dyebath water in which the dyestuff has been cooked out. The dyestuff itself is often subsequently strained off and discarded.

dyestuff animal, mineral, or vegetable (plant) matter from which a dye is made

fastness property of plant-dyed fibre which enables it to resist fading upon exposure to light and water (washing)

fermentation processing of a dyestuff (usually umbilicate lichens) in water, ammonia, and oxygen for several weeks or longer, until the mixture 'works,' like homemade wine

fibre thread or strand; a yarn or cloth composed of threads, either natural (wool, cotton, linen) or man-made (nylon, rayon)

fleece raw wool, unprocessed and unspun. 'In the grease' refers to unwashed fleece, and 'scoured' to fleece that has been washed. Wool is fleece until it is spun into a continuous thread.

fugitive quality of a pigment which renders it non-permanent. A fugitive colour (see beet, p 102) will fade upon exposure of the dyed fibre to light or washing; a non-fast colour

indigenous native to a region or locale; naturally produced or born in a specific place

levelling term used to describe the addition of sodium sulphate (Glauber's salts) to a dyebath to act as an agent which evens the colour of the bath

lichen form of plant life composed of two organisms: an algae and a fungus; not to be confused with moss. In Cape Breton, dye lichens are referred to as 'crottle' or in the Scottish form, 'crottal' (Seonaid Robertson, *Dyes from Plants* 104)

liquor strained-off dyebath; the liquid in which the dyestuff has been cooked out

macerate to reduce to a soft mass by soaking in a liquid

mordant additive, most often in the form of chemical salts (alum, iron, tin); helps make the dye fast and affects its colour by brightening, darkening, or otherwise changing the dye colour obtained

natural dye fibre dye from animal, mineral, or vegetable matter; synonymous with organic, plant, and vegetable dyeing

orchil substance present in certain lichens (*Parmelia, Umbilicaria*) which, when fermented with water, ammonia, and oxygen, produces red

pH pH value of a solution refers to its degree of acidity or alkalinity. The pH scale ranges from 0 to 14, with 7 as the mid-point (neutral).

ply strand of a yarn. Three-ply refers to a yarn made up of three strands of any weight or thickness.

rinse to wash in water without soap; to remove from dyed fibre any trace of the dye liquor or mordants which may remain after dyeing

scour to wash in warm soapy water until all foreign material and excess natural grease are removed; particularly applied to the preparation of fleece for spinning and dyeing

simmer liquid at a temperature of 95°C and 200°F; temperature at which the dyebath is processed when the fibre and mordants have been added

skein continuous, circular hank of yarn; usually 4 oz (114 g) in weight, if commercially wound

soak out to wet yarn thoroughly; to aid the extraction of pigment from a dyestuff by soaking it in water for several hours or longer

substantive dyestuff that will impart colour to a fibre without the use of mordants. Most boiling water lichens are substantive.

wool natural fibre, from sheep, in any form. The term applies equally to raw fleece and finished cloth; pure wool has no other additives, but wool yarn may or may not have sizing applied to it when it is factory-spun or dyed.

yarn spun fibre, of any content; available in various sizes and weights, from one-ply (singles) to two; three; four; or five-ply

2

Equipment

The amount of equipment purchased for making dyes depends on the dyer's seriousness. Many pounds of fibre can be successfully dyed using utensils found around the home or studio. Nothing used for plant dyeing, however, should ever be used subsequently in the preparation or cooking of food. This means using old pots and pans or buying new ones. For dyers whose long-term interest warrants it, buying quality equipment at the outset is a good investment. A stainless steel pot may cost four or five times the price of an enamel one, but it will last for years, whereas the enamel pot will rust long before and have to be replaced. Dyers may also discover that it is difficult to obtain large enamel canners out of season.

Dyers using hotplates will find it takes longer for the dyepots to heat than if a kitchen range is used as a source of heat. Wood-burning stoves are excellent and cheaper to operate than electric ranges (see appendix, p 223). However, it is quite safe to use the kitchen stove for dyeing if some precautions are taken (see safety precautions, p 67). If there is an exhaust fan or hood, use it to dissipate the fumes from the dyepot. Otherwise, work with an open window, even in the winter months. Although many dyebaths are alarmingly disagreeable in smell (seaweed, for instance), most are not toxic. All mixtures are potentially hazardous, however, and may produce toxic fumes. Using rubber gloves, keeping a window open, and cleaning up carefully afterwards should become a ritual for dyers. Teachers are advised to work with non-poisonous household mordants (see classroom techniques, p 80).

EQUIPMENT LIST FOR BEGINNING DYERS

– Enamel canning pot, with lid. As the same pots are used for mordanting and dyeing, one will do, preferably the size that holds 8 canning jars in the rack.

– plastic ice cream containers, 1-gallon or 4-litre size, or a plastic bucket (for soaking out dyestuffs and holding strained-off dye liquor)
– wooden stick, dowel, or piece of driftwood for stirring
– cheesecloth, old nylons, or plastic colander for straining off the dyestuff after it has been cooked out
– plastic measuring spoons and glass measuring cup (old style, 32-ounce size; new type, 1-litre size or 35.2 fluid ounces)
– rubber gloves, as heavy as possible
– old newspaper, rags, and clean-up supplies
– the collected dyestuff; several gallons of rain or tap water; one or more of the following household mordants: baking soda, common salt, drugstore alum (see alum, p 23); urine; vinegar; cream of tartar

RECOMMENDED EQUIPMENT FOR ADVANCED DYERS

– 1 stainless steel stock pot with lid; 2 or 3 canners with lids; 1 separate 'iron' pot (for iron mordanting and dyeing), which may be cast iron or enamel; 1 aluminum pot (for pot-as-mordant dyeing); 1 tin pot (for pot-as-mordant dyeing); a copper-lined tin wash boiler, or an antique copper or brass pot. Because most serious dyers give workshops and demonstrations, it is important to have on hand a variety of equipment suited to these situations.
– several plastic pails with handles and pouring spouts
– wooden stirrers; plexiglass stirrer for indigo dyeing
– several plastic colanders which, if possible, will fit exactly on top of the plastic pails
– plastic measuring spoons; several glass measuring cups (old style, 32 fluid ounces; new type, 1 litre)
– several pairs of lined rubber gloves
– clean-up equipment
– common mordants and, in addition: alum, blue vitriol, chrome, hydros, iron, lye, tin, cream of tartar, Glauber's salts
– scales (baby scales are better than diet or kitchen scales as the latter tend to be flimsy in construction)
– thermometer (in Celsius and Fahrenheit)
– litmus paper to test pH
– shears or pruners, preferably with non-iron blades
– blender, or mortar and pestle (for grinding umbilicate lichens)
– folding clothes rack (wooden, or rubber-covered aluminum)
– looseleaf binder for yarn samples and notes

DYEPOTS, OLD AND NEW

Buying quality equipment means it will last for many years and prove less expensive in the long term. A stainless steel 'stock pot,' such as those used in professional kitchens, is invaluable to the serious dyer, as are a variety of antique vessels. Although the initial cost of a stock pot is high ($100), it costs less than a much smaller antique pot. Most stock pots hold at least 10 gallons (45.46 l) and, if well cared for, last for twenty to thirty years. Brass, copper, iron, and tin pots occasionally turn up at country auctions or dealers' shops. Regrettably, they are now almost prohibitive in price, affordable only by either the very rich or the very dedicated. Still, the challenge presented in using such a pot can be exciting.

Top prices are paid for vessels in prime condition. That means they are clean on the inside, polished outside, free from large dents, have no leaks, and have a functioning handle. They may or may not have lids. Premium-quality antique pots are more desirable to the collector than to the dyer. The overall condition of a pot need not be perfect, and a lid from another vessel can be used for a cover. Cast iron hearth pots are excellent for dyeing, but professional cleaning of the interior may be required before they can be used. Copper and tin sap buckets make good dyepots, but the most suitable vessels in the antique category are brass porridge pots and copper jelly pans. A small porridge pot can cost as much as $80 to $90, and a copper jelly pan with a capacity of 2 gallons (9.092 l) will fetch upwards of $125 at most auctions. Dyers who want an antique pot are advised to remember that at auctions, anything goes. A deeply encrusted iron pot may be impossible to clean. However, a reputable dealer may be able to locate a good pot, and there is the advantage of knowing that, when purchased, such a vessel will probably not require repairs.

REPAIRING ANTIQUE POTS

A jeweller or silversmith will usually be able to fix brass and copperware, and possibly tin, depending upon the character of the tin. No severely rusted metal pot should ever be considered useful for dyeing, nor is it worth having repaired. Some blacksmiths will fix cast iron pots, although most prefer not to. However, with the increase in the number of ironsmiths in northeastern North America one can take cast iron pots for repairs. Handles can be replaced and useful devices such as trivets for hot dyepots can be fashioned.

THE POT CONTROVERSY

During the past forty or so years there has been a strong feeling among textile dyers that to use anything other than an enamel or stainless steel pot would be detrimental to the quality of dyed goods. This belief is rightly based on the fact that all metals other than enamel or stainless steel react chemically with the dye and mordants in the pot and thereby affect the resulting colour. The controversy regarding what type of dye-pot to use is still an issue when dyers meet and work together. However, most now take the sensible view that it does not matter what metal a dye vessel is made of, as long as the dyer understands how each will affect the dyebath itself. Pots of copper, brass, aluminum, iron, and tin will alter the colour of a dyebath, as will the addition of a mordant. It is essential to understand this when reading dyeing reference material, as few authorities agree on the type of pot to use. Despite individual preferences, the consensus seems to be in favour of stainless steel or enamel.

Edward Worst, writing early in this century, advised his readers that 'copper kettles are best' ('Dyes and Dyeing' 5). Rita J. Adrosko (*Natural Dyes and Home Dyeing* 65), and Margaret S. Furry and Bess M. Viemont (*Home Dyeing with Natural Dyes* 4) concur with that opinion, and suggest that enamel is equally satisfactory. Elsie Davenport (*Your Yarn Dyeing* 41) and Seonaid Robertson (*Dyes from Plants* 13) favour stainless steel, followed by enamel, while Mary Frances Davidson (*The Dye-Pot* 4) recommends pots made from glass or granite.

Enamel pots are available at department and hardware stores. Those with missing lids are a good buy for the dyer who already has enough covers. A 5-gallon (22.73 l) canning pot sells for around $10, with larger ones higher in price ($15–$20). Avoid buying blanchers unless you can use the extra steaming unit in your kitchen, as they are more expensive than regular canning kettles. Aluminum pots are sold nearly everywhere one shops for kitchen utensils, and a few rural and specialty hardware stores still sell tin-lined copper wash boilers for $30 to $40. Stainless steel stock pots are available only through specialty shops that cater to the serious cook. Large sizes are usually not in stock, and a deposit is required before your order is taken. War surplus stores carry inexpensive tin pots, and, occasionally, aluminum or enamel kettles.

Experience, budget, and availability help the dyer determine which type of pot is best suited to particular needs. All dyers will find using a different pot from their regular one an interesting challenge, especially when adding these newly dyed samples to their file. A completely different range of colours may be obtained. When these are juxtaposed with previously dyed samples, the dyer can immediately see how his or her colour vocabulary has been diversified and expanded.

OTHER UTENSILS

As is the case with dyepots, dyers still disagree as to what should be used for stirring the dyebath and removing the yarn. Some insist on wooden dowels or spoons, with one such stirrer for each mordant used (*Journal of the Chicago Horticultural Society* 20). Violetta Thurston recommends using glass rod stirrers, but plexiglass is cheaper (*The Use of Vegetable Dyes* 7). Glass rods chip easily and can ruin expensive rubber gloves. Wooden dowels and spoons are inexpensive and driftwood is free. Some dyers use old enamel curtain rods for stirring. Although it is widely believed that pigment will build up on a stirrer and thereby spoil subsequent light-coloured dyebaths, I have not found this to be the case, with two exceptions as noted below. The pigment appears to penetrate the wooden stirrer fully and remain there permanently. It does not rub off, even on a light-coloured cloth. An exception would be made in the case of indigo, woad, or the umbilicate lichens. These baths require a separate stirrer used only for dyeing with these dyeplants. In earlier times in Nova Scotia, it seems that almost every household had a special indigo stirrer, 'made from ash and smoothed of every splinter. This was called the "indigo stick"...' (Margaret MacPhail, *The Bride of Loch Bras d'Or* 21).

Plastic colanders for straining off the dye liquor are the most convenient to use if they fit exactly on top of the plastic pail or bucket used to hold the liquor. This results in the dyer's hands being free to hold the hot dyepot and better control the stream of liquid as it is poured off. If using a cloth for straining, wet it first. This will prevent the cloth itself from absorbing too much of the valuable dye liquor. Tie the cloth securely around the top of the pail with string, or clip it in place with clothes-pins.

Large glass measuring cups (old style, 32 oz; new type, 1 l) enable the dyer to dissolve mordants fully which are in powder or crystal form, and prevent effervescent substances such as baking soda from spilling over the top. It is also much easier to rest a wooden stirrer safely in a large cup than in a small one.

Heavy-duty rubber gloves are essential for all dyers. They not only protect the dyer's hands, but health as well. Poisonous substances, such as mordants, must not come into direct contact with the skin, especially after they are in solution with water. Surgical gloves are generally unsatisfactory because they do not protect the dyer's hands from the hot dyebath, and can tear as they are being removed. Very thick gloves of the type used by kitchen workers in institutions and restaurants are so heavy they prevent the dyer from being able to hold the handle of a measuring cup or a measuring spoon. Also, because they are so much larger than the hand, there is the danger of liquid flowing in over the top of the glove when it is submerged in the dyepot. The ideal glove is made of blue rubber, and has a white knitted cotton lining. These are expensive, but two pairs will last most dyers a year. (Use a separate pair for dyeing with indigo or woad.)

Scales are required by serious dyers. The old-fashioned balance scales often seen in country stores are ideal, although quite expensive. They have large pans, which makes it extremely easy to weigh dyestuffs and fibres. Diet scales featuring metric measurements are available at most department stores, as are scales for weighing mail. Look for the largest scale you can afford, preferably one which will weigh amounts up to several kilos.

A folding wooden clothes-rack allows the dyer to dry many skeins at one time. Metal drying racks are satisfactory as long as the 'arms' are plastic-covered. Hanging wet, freshly dyed skeins over a steel or aluminum rod could cause the yarn to change colour where it came into contact with the metal. Light-weight aluminum sweater racks which are suspended over the bathtub make good drying racks if they are first covered with a large sheet of plastic. Racks can be purchased at most hardware and department stores, as well as at auctions.

Equipment for dyeing is as personal a matter as decisions made regarding what dyestuffs to collect. Each dyer develops methods and then acquires appropriate utensils for them. What equipment a dyer buys depends upon location, budget, and degree of interest. Suggested utensils are just that – suggestions. Be flexible and improvise where necessary.

CARE OF EQUIPMENT

No matter what a dyer uses for dyeing, equipment must be properly cared for in order for each dyebath to be potentially successful. Dyers who use enamel pots will find them extremely difficult to keep clean and rust-free. After each dyebath, the pot should be thoroughly scrubbed using soap and a rough cloth. Then it is rinsed with warm water. Next, fill the pot half-full with warm water and add to it one cup of chlorine bleach. Add more water to the pot until it almost reaches the top. Allow this solution to sit for half an hour. Pour it off and dry the pot with paper towels. If any colour shows on the towels, repeat the bleaching. If not, leave the pot upside down in a warm place to dry. Prop one edge of the pot so that air can circulate underneath it. Never put the cover back on a pot that has just been washed. Do not store enamel pots one inside the other. Stainless steel pots need only a soapy rinse to clean them, but they too should be stored uncovered, and upside down, with one side propped. Antique copper and brass pots should be washed immediately after using with warm, soapy water, followed by a cleansing with a special brass or copper cleaner. However, after using such a cleaning agent, the pot should be washed again with warm, soapy water. A mild bleach solution (one-half cup of bleach to a pot full of water) will remove stubborn stains. Tin-lined wash boilers benefit from bleaching and a subsequent water and vinegar rinse (one cup of vinegar to one gallon, 4 litres, of water); after this, the pot may be washed with warm soapy water and then dried with paper towels. It is essential to dry all dye vessels carefully in order to prevent rusting. Dyers who wish to soak dyestuffs for long periods of time, or leave the dyed fibre in the dyebath overnight, are advised to pour the mixture into a plastic basin or pail.

All glass utensils, such as measuring cups and screw-top jars, should be cleaned after each use with water and bleach. Do not attempt to remove stains from glassware by scouring, as the abrasion will pit the surface of the glass and cause it to stain even more another time.

All dyeing equipment should be labelled as such. Marking pens can be used on metal, glass, plastic, and wooden surfaces. Even pot lids should be designated as dyeing equipment if there is any danger that another member of the household might mistake the cover for one used for a cooking pot. Read safety precautions, page 67, before mordanting and dyeing for the first time.

3

Where to go and what to look for

Beginning dyers often have trouble locating suitable dyestuffs. What should you look for? That depends upon several factors, including an understanding of those plants which are easy to use, the availability of transportation, the locale, and the time of year. For instance, a novice dyer living in a city would be ill advised to search there for buttercups in May. Buttercups bloom in June, and prefer hay fields and country pastures. On the other hand, the same city dyer might well locate rhododendron leaves at the local park from a bush that suffered winter kill. Know what you want to collect first, and how to use it. Learn all you can about plants, and their habits of growth. Then go out and look.

URBAN COLLECTING: SUMMER

Every park or horticultural garden of any size has a superintendent whose aim it is to maintain the grass, trees, shrubs, and flowers at their peak condition. The superintendent also has a staff, and getting to know them means a dyer will never want for dyestuffs. But before you start collecting at a park, introduce yourself to the head person, who will then refer you to someone else, perhaps the 'flower man,' or the 'shrub man,' who will see that you get whatever is being discarded. Disease and severe winter conditions kill plant life, and everything that is thrown away is useful to the dyer. Oak, maple, beech, sumac, walnut, ash, birch, elm, linden, locust – all are valuable for their leaves and bark. Black walnut hulls, acorns from oaks, and the nuts of the butternut all yield excellent dyes. The leaves of many ornamental shrubs such as rhododendron yield fine colours, as do those of most fruit trees, including apple, crab-apple, pear, peach, and cherry. Lilac blooms may be picked after they fade and used for a dye. Most gardeners bemoan the fact that they have to pick off wilted petunias, pansies, and marigolds to prolong the season of bloom, but what is useless for ornamentation is usually perfect for dyeing. Park staff prefer to do the picking, however. This discourages onlookers from following your example and raiding the flower beds. Ask the staff if they wish you to leave a container with your name and phone number on it. Then, when it has been filled, they can call you. Most park superintendents are extremely helpful and go out of their way to assist dyers, so co-operate with them by taking their advice and not asking for special favours unless the situation demands it.

Supermarkets are filled with fresh produce in the summer, as are outdoor urban markets. Rhubarb leaves and carrot tops are two examples of plant material considered to be waste, yet both yield outstanding dyes. Speak to the produce manager, explain what you want, and arrange a mutually convenient time for pick-up. Offer to leave a container with your name and phone number. The leaves of lettuce discarded each day in a supermarket would fill several cartons.

URBAN COLLECTING: WINTER

Aside from park superintendents, the city dyer should also cultivate friendship with the local florist. Cuttings, stems, leaves, and wilted blooms are usually thrown out at the end of each day. The best time to collect from the florist is immediately after a social event such as a dance or social benefit. Another winter source of dyestuffs for the city-dweller is the power company. Its trucks make the rounds of city streets following storms to remove broken or damaged tree limbs. Phone the line maintenance supervisor and arrange to meet the trucks, or ask permission to intercept them en route. All types of valuable shade tree bark may be obtained in this manner. Fireplace owners always have wood on hand, and many do not mind parting with the bark from it. Apple is especially good for dyeing, as is birch. Wood ashes make an excellent dye, although their collection from several households is messy and time-consuming. Have friends save tea bags or coffee grounds for you. Most cities have at least one vegetable packing warehouse whose floors are littered with a variety of dyestuffs, including onion skins. These warehouses employ people for the sole purpose of sweeping up the skins, so your removing garbage bags full of them will be welcomed. Phone ahead to determine the best time to visit, and take along a broom and dustpan. The staff is usually so impressed with your neatness that they will welcome you back.

RURAL COLLECTING: SUMMER

It would be virtually impossible to list all potential dyestuffs available to dyers who live in rural areas. Suffice it to say that almost everything that grows will give a dye. Some plants yield better colours than others, in the sense that they are more fast and perhaps easier to use. Among the outstanding summer dyestuffs, listed in order of their appearance, are: dock, dandelion, lupins, daisies, pink clover, fruit tree blossoms and new leaves, wild rose leaves, rhubarb leaves, carrot tops, lettuce, Swiss chard, radish tops, St John's wort, evening primrose, goldenrod, raspberry and blackberry leaves, wild asters, mushrooms, wild mustard, rose hips. The list is endless. Weeds are a source of dye, as are all wildflowers. Leaves of various types of trees give a dye, as do all garden flowers. Refuse from the home garen is useful to the dyer; everything from spinach that has gone to seed to tomato vines may be used successfully in a dyebath. Beets give a colour that is fugitive (see p 102), but most beginning dyers try them anyway. Mushroom collecting, from May through late September, is dangerous for dyers who cannot identify poisonous species. Even handling these with gloves can be harmful. Read about mushrooms, page 180, before using them for a dye. If you do not know most weeds and wildflowers by sight, ask neighbours to point them out to you. Most farmers know the common and vernacular nomenclature for almost every plant that grows on their property. Take the benefit of such a wealth of experience, and learn about plants.

RURAL COLLECTING: WINTER

Dyers who live in the country do not have access to produce departments in supermarkets, but they are able to collect dyestuffs outdoors long after the season ends for city-dwellers. Wild mustard and wild aster are only two of a variety of wildflowers which bloom into November in many parts of the northeast. Rural areas are also the locale for harvest suppers, church bean bakes, and similar community events. Hundreds of tea bags, pounds of coffee grounds, and often crates of vegetable peelings are available to the dyer willing to collect them. Lichens may be collected in the winter, although the colours they give are not as strong as if they are picked during summer (see lichens, p 164). But dyers are urged to STAY OUT OF THE WOODS until after the hunting season is over. This date varies from year to year, with the moose season shorter than the deer season. Check with the Department of Lands and Forests in your region before hiking in the fall. One advantage of collecting lichens in the winter, particularly for samples to file away for reference later, is that these subtly-coloured fungi stand out against the drab off-season landscape and are much easier to discern. For dyeing, take only those species which appear to be prolific; otherwise collect only a small sample for identification. Look for deadfalls in the woods, and collect bark from these, rather than from living trees. At the base of most paper birches you will find a small heap of bark. The only wood that may be freely used at any season of the year, in the form of twigs, is the ubiquitous alder. This clumpy small tree is a nuisance to farmers, and no one will mind your removing as much of it as you wish. Ask first. The tips of most evergreen trees yield a fine colour, but collect only a few from each tree in locations where there is an established stand. Collect the leftover Christmas trees from the local school, or strip the branches of their needles in the school yard and carry them home in pails. Ask neighbours to save onion skins, especially at pickling time. These can be kept for dyeing for many months, if they are stored in a plastic bag with holes in it to allow air to circulate and prevent mustiness. Some dyers use Hallowe'en pumpkins for dyeing (the accumulated soot acts as a mordant). As is always the case, the use of food for dyeing is a matter left up to the individual dyer.

COLLECTION EQUIPMENT

Carry more containers with you when collecting dyestuffs than you expect to need. Far too often dyers find themselves without a spare bag or pail when a particularly desirable dyestuff is growing in abundance all around them. Also, it is easier to have a separate container for each dyestuff collected than to toss all into the same bag. Sorting plants out, before starting the dyeing, is tedious and time-consuming.

Shears, scissors, a knife, and a spoon are essential equipment when collecting tough-stemmed or shrubby plants, weeds with rubbery stems, and lichens. Dyers who pull off apple tree leaves will have yellow hands within fifteen minutes, so wearing gloves is advisable if you'd rather have the pigment on the fibre that is to be dyed. Gloves are also necessary when picking rose hips, wild rose leaves, or nettles.

Large-bloomed weeds and wildflowers, such as goldenrod, are best collected in baskets or cardboard cartons. If stored in plastic bags, even for a short time, the heads give off sufficient moisture to turn the whole mass into a sodden heap. On a humid day, a plastic bag full of leaves or mushrooms can become a mildewed mess if the dyer places the bag in the trunk of the car and takes several hours to reach home. If dyestuffs must be stored in plastic bags for any length of time, leave the tops undone, and avoid placing them in car trunks. Cardboard boxes are excellent for use as collection containers, as are bread bags with holes punched in them. One can carry many such bags easily by tying them together and slinging them from belt loops. Remember that once you have walked two miles to reach a dyestuff, it pays to have an idea of just how you will cart it home. Dyers who have collected wet seaweed in any amount will appreciate the significance of that statement. For testing the presence of orchil in a lichen (see p 172), dyers should carry a small bottle of household bleach and a razor blade. The blade can be taped along one edge of a pocket or carried in a small aspirin tin.

No matter what the season, dyers who collect on a regular basis usually wear rubber boots and heavy socks. This is advisable in the woods, even during summer. Hiking boots are just as good, as they provide the traction often needed when climbing a steep slope or even a tree to collect a lichen. As most lichens are easier to gather on a wet day or when there is a heavy fog, rubber rainwear is essential. Wear gloves, just in case. You may come across something with thorns. If unfamiliar with the woods, have a compass and a companion with you. Tell another person where you will be. And have as many field guide books as you can stuff into your pockets. Never collect mushrooms without at least one field guide to refer to on the spot.

Dyers who dislike chance meetings with porcupines, reptiles, or bears will discover the joys of collecting in November. Most bears hibernate, which makes the winter months quite safe. I have not found porcupines to be anything less than charming, but there are people who find their presence frightening. Snakes are rarely in evidence before May or after late September. It always helps, if you must collect alone, to take along the family dog. Have a leash with you, just in case. Remember to be honest with yourself before taking off into the woods: if you are not familiar with the territory, then don't go. Join a hiking or orienteering group to develop skills in using a compass. Don't fight nature – go with it. Collecting dyestuffs provides many dyers with a marvelous opportunity to see wildlife, an opportunity they might otherwise miss.

4

Mordants: What they are and how to use them

There is more confusion regarding the use of mordants than with any other aspect of plant dyeing. Mordants are simply metallic or mineral salts which, when added to the dyebath, enhance, intensify, or change the colour of the dyebath and make the resulting shade more fast to light and washing.

SUBSTANTIVE AND ADJECTIVE DYESTUFFS

There are two types of dyestuffs: substantive and adjective. A substantive dyeplant or lichen is one which will impart colour to the fibre being dyed without the use of a mordant. *Lobaria pulmonaria* and other boiling water lichens are substantive (see lichens, p 164), as are tree bark and twigs. The orchil-yielding *Umbilicaria* species do not require mordants in the dyepot, although ammonia is used during the fermentation period which precedes dyeing. However, while mordants are not a necessary additive with substantive dyestuffs, the use of chemicals does greatly extend their colour potential. Mordants may also make *Umbilicaria*-dyed fibres more fast to light.

An adjective dyestuff requires the addition of a mordant to the dyebath to intensify the colour obtained and make it permanent. Most common garden and wild flowers are adjective, as are the majority of weeds, berries, mushrooms, and vegetables.

EFFECT OF MORDANTS ON AN ONION BATH

Although onion skins are substantive, few other dyestuffs respond as dramatically when mordants are added to their bath. The profound influence which a mordant effects is described in Table 1. The mordants chosen to illustrate how a dyebath is altered and changes by their addition are chrome and iron (see pages 25 and 26).

Table 1: Effect of mordants on an onion bath

	no mordants onion bath	chrome onion bath	iron onion bath
weak	soft yellow	bright yellow-orange	tan
medium	medium yellow	bright orange	khaki
strong	old gold	dark brown	olive-green

weak bath: 1 part onion skins to 2 parts fibre (by weight)
medium bath: equal parts onion skins and fibre
strong bath: 2 parts onion skins to 1 part fibre

The colours described in Table 1 were obtained using 1 pound of fibre (453 g) and the amount of each mordant as stated on page 29. The onions were home-grown and had rusty brown skins. The pale yellow commercial variety would yield generally lighter shades. For instance, in a strong bath, with a chrome mordant, the resulting colour would perhaps be rust instead of dark brown.

Lacking the necessary chemicals for mordanting, dyers often use the dyepot itself as a mordant (see p 38). Referring to Table 1, a tin pot would result in bright, sharp colours without the addition of any mordant. A copper dyepot would impart a green tone to the colours obtained, while an aluminum pot would produce much the same result as the addition of alum. Using a brass pot gives tones of bronze, gold, and ochre.

HOUSEHOLD MORDANTS

No one mordant is more effective than any other; nor is it 'better' or 'worse.' Among the common household mordants, dyers generally choose to use those immediately on hand whose effects they have come to understand or anticipate. What is attractive to one person is not necessarily so to another, and this is certainly the case when working with colour. Do keep in mind that even plain beige and tan can be extremely useful when planning a craft project.

alum (see p 25). There are two varieties of alum, which are chemically different. 'Household' alum is aluminum ammonium sulphate, a non-poisonous powder often used in first aid treatments. Although this drugstore variety does not produce the same shades as the other, poisonous, alum, it can be useful as a mordant. This is the alum recommended for use by persons demonstrating dyeing to children or working in the classroom.

ammonia Ammonia is most often used as an additive when fermenting orchil-yielding umbilicate lichens. An alkali, it must be used sparingly in the dyebath or the quality of the fibre will be impaired. A small amount added to a yellow bath (such as goldenrod) will result in a yellowish-green colour.

baking soda Bicarbonate of soda is also an alkali which is useful in changing yellow dyebaths to yellow-green. The most effervescent of all mordants, baking soda must be used with care so that the dyebath does not bubble over. Add soda to the dyepot only when the liquid in the pot is less than half-way up the side, and never cover the pot after the addition. Used in this way, and always at a temperature which is below the simmering point (see p 8), baking soda can be an interesting mordant to use in the classroom and at demonstrations.

cream of tartar Potassium bitartrate, cream of tartar, is used as an additive to alum and other mordant baths. It softens the effects of harsh mordants such as tin and iron, and increases the acid value of a bath. Although expensive when purchased at supermarkets, cream of tartar is much less so when bought from chemical suppliers.

salt Sodium chloride is common table salt. There are several forms: sea salt (purchased from health food suppliers); coarse pickling salt (from supermarkets), and ordinary iodized table salt. The addition of salt to a dyebath softens the colours and slightly retards the rate at which the fibre absorbs the dye. It is used to 'draw' the colour from certain dyestuffs, namely flowers, barks, lichens, and roots, and is sometimes added to these when they are soaking out. A final hot water and salt rinse is beneficial when rinsing out some colours after the dyeing. This aids in the fastness of these shades and prevents too much colour 'rubbing off' after the yarns are dry.

urine Human or animal urine is suitable for use as a mordant, and improves with 'ageing.' It is available in urban areas from hospitals and laboratories. Urine may be used instead of ammonia to produce approximately the same results in the fermentation of *Umbilicaria*, but it is weaker than ammonia, so more must be used. Traditionally used in indigo dyeing in Cape Breton, MacPhail advises that 'urine was only collected in winter time. Animal urine ... was easier to process than human urine' (*The Pride of Loch Bras d'Or* 19). The odour is such that only a few dedicated dyers use it today, in demonstration or experimentation.

vinegar Vinegar is actually a weak solution of acetic acid. When added to the dyebath it helps to soften hard water and somewhat brightens and darkens colours. It also neutralizes an alkaline bath and is used as a final rinse to neutralize fibres dyed in indigo (p 157). The addition of vinegar to an offensive-smelling dyebath somewhat reduces the odour. White vinegar is preferred, but when dyeing rich browns try using cider or malt vinegars.

washing soda Hydrated sodium carbonate or washing soda is also called sal soda. An alkaline substance, it is very strong and should be used with care in order not to damage the fibre. When added to an *Umbilicaria* pot, washing soda swings the red colour towards bluish-purple. Stronger than baking soda, washing soda should be thoroughly dissolved before adding it to the dyebath and the temperature of the pot must then be kept below a simmer. Alkalines may impair the quality of fibre when in contact with it above temperatures exceeding 200°F or 95°C.

CHEMICAL MORDANTS

The mordants listed below are chemicals which must be used with care. Some are poisonous and most are skin and throat irritives. All may be toxic when combined with other chemicals in solution with water. Read pages 31 and 67 before using these mordants. Keep them out of the reach of children. Label all containers and wear gloves when handling chemicals.

alum (See alum p 23, household mordants). Aluminum potassium sulphate, or what Douglas Leechman (*Vegetable Dyes from North American Plants* 14) calls 'potash alum' is the alum used by dyers. A poison, it may be purchased in bulk from chemical suppliers. The use of cream of tartar with alum is recommended, as it softens the effects of the alum. Using too much alum will result in the dyed fibre feeling gummy and harsh. Alum causes yellow baths to take on a greater intensity and brightens most light colours.

blue vitriol Traditionally called 'bluestone,' blue vitriol is useful in obtaining greens from yellow baths. Copper sulphate, in solution, is a weak form of sulphuric acid and is poisonous. Once a yellow bath is removed from the heat, often the addition of blue vitriol and iron will turn the bath a good green without further processing. It is also useful in obtaining greens by top-dyeing.

chrome Potassium dichromate is a bright orange chemical often thought to be light-sensitive when in solution with water, a fact that some references refute (*Journal of the Chicago Horticultural Society* 19). My own tests indicate that chrome-mordanted and -dyed fibres processed away from direct light are somewhat darker and deeper in hue than if processed in the light. See pages 33 and 36. Dyers are advised to conduct their own experiments in using chrome, and decide for themselves which procedure to follow. Chrome added to a yellow bath produces startling results, turning it orange or rust. A poison, chrome is no more difficult to use than other mordants. If you wish to process it in the dark, work with the blinds drawn or just at dusk. The addition of chrome to a bath makes the fibre feel soft and silky, and imparts as well a very attractive lustre.

Glauber's salts Sodium sulphate is a purgative used in dyeing as a levelling agent. The addition of these salts to a dyebath ensures even dyeing. Glauber's salts may darken colours somewhat. It is also used to extract pigment from barks, lichens, and roots, much as ordinary salt is used to 'draw out' colour, although Glauber's salts are stronger than the household variety. Although some references recommend the addition of Glauber's salts to every dyebath (Alma Lesch, *Vegetable Dyeing* 26), this results in the dyer producing a limited range of shades, and all with that quality of colour which Glauber's salts produces. However, it is extremely useful to add these salts when attempting to level dyebaths.

iron Ferrous sulphate is sometimes referred to as 'copperas,' which is its traditional name, but the use of that term confuses new dyers unfamiliar with the terminology. Iron 'saddens' colours; in other words, it drabs them, producing khaki and avocado greens from strong yellow-green dyebaths such as goldenrod and rhododendron leaves. It will also turn a pale beige or tan to a good grey and make excellent medium greens from a lettuce bath when combined with blue vitriol. Iron is poisonous. It is important not to use more than the specified amount (p 29) as too much will make the fibre harsh. Keep a separate pot for mordanting and dyeing with iron, as even a little residue from it in the pot will darken and drab subsequent alum, chrome, and tin baths. If new dyers have a single dyepot, they may find it preferable to use an iron substitute (see p 27).

tin Stannous chloride is perhaps the most difficult mordant to use successfully, as tin-mordanted and -dyed fibre often streaks and feels 'crimpy' and harsh. Great care must be taken to follow a few basic procedures. These describe the mordanting process for all mordants, but only with tin is a dyer's carelessness so obvious after the fibre is dry. First, thoroughly wet the fibre to be mordanted or dyed; make certain the tin crystals are thoroughly dissolved in boiling water; turn the fibre over in the pot every two or three minutes; set a stainless steel rack in the bottom of the dyepot so that the fibre will not touch the 'hot spots' on the pot's bottom; keep the temperature below a simmer (200°F, 95°C); and rinse after dyeing in several baths of water, each slightly cooler than the last. Tin is a popular mordant, as it produces the most dramatic shades of any chemical additive. It turns pale yellows to brilliant yellows, yellow-gold to bright orange, beige to russet. One of the finest yellows I have seen was produced by adding tin to a bath made from fresh hemlock tips. Subsequent dyeing produced the exact shade time and time again, giving the dyer several pounds of yarn precisely the same colour. Tin is poisonous, and the most costly of all the chemical mordants. Use sparingly, as a little goes a long way. New dyers will find that 2 oz (57 g) is sufficient to start with.

OLD AND MODERN MORDANT SUBSTITUTES

Eighteenth- and nineteenth-century home dyers in the eastern North American colonies used a variety of readily available kitchen and boudoir products as mordants. Wood ashes, blood, urine, lye, tree galls, and clay were popular. Blood was readily available from slaughterhouses, and urine was collected daily by dumping the chamber pots into a collection barrel stored in the cellar (MacPhail, *The Bride of Loch Bras d'Or* 20). Wood ashes were used both as a mordant and a dye, for greys and blacks. Apparently they are still used as a dye in the Hebrides, on the Isle of Harris. Using no mordant, Mrs Marian Campbell of Plockropool obtained a fine golden brown and rust from peat ash, or 'soot' as she called it. The water used in the processing may well have determined the brown result rather than the more usual grey or black (information collected by Lauren Dougall Vaughn, Windsor, NS).

The eastern Indians used the acid juices of crab-apple and other indigenous fruit to set their dyes (Douglas Leechman, 'Aboriginal Dyes in Canada' 71), and a contemporary Eskasoni dyer and basket-weaver, Margaret Johnson, cooks out her alder brown in an aluminum pot, which acts as a mordant for the wood dye. Basket-maker Edith Clayton of Preston also cooks alder in an aluminum pot. Mrs Clayton is not consciously using the pot as a mordant. The same is true of Mrs Johnson. Mrs Clayton also adds baking powder 'sometimes' to the alder bath, and thinks spring-cut wood gives the best colours. Alder itself is used as a mordant in Ireland, with dyers taking advantage of the natural tannin it contains, as reported by Halifax dyer Joan Doherty after a trip to Ireland in 1975. Alder is also used sometimes in lichen dyeing in County Kerry, serving as a mordant (Lillias Mitchell, *The Wonderful Work of the Weaver* 23).

Sumac was, and still is, used for a mordant (see p 210). The fresh twigs are cooked out and the liquid strained off. This may then be added directly to the dyebath. Oldtime dyers often put the sumac, the dyestuff, and the fibre all in the pot at once. Tree galls, which are large, roundish distortions seen on limbs of a variety of species, were a popular mordanting agent. They were first dried in the sun, and then ground into a powder. Iron nails and old horseshoes may be used instead of an iron mordant, and a discarded tin can serve a similar purpose to the chemical stannous chloride. Copper scouring pads or a small brass rod may be added to the dyepot to produce interesting effects instead of using a vessel of either metal. A piece of aluminum foil reacts chemically in the dyepot, as would using an aluminum pot. The variations are many: use your imagination. It is often just as satisfying to use a substitute as to use a chemical mordant.

HOW MUCH TO USE

Dyers rarely agree on the amount of a mordant to use. The measurements given here are underestimated rather than too generous. This is because using too much of any chemical mordant will impair the quality of the mordanted and dyed fibre. Each dyer usually discovers the correct proportions by experimentation. Owing to the different soil conditions in which dyeplants grow, variations in climate, and the water used in mordanting and dyeing, no two dyers obtain exactly the same results even from an identical amount of the same mordant in the same dyebath. Although dyers are often able to reproduce a colour following their own documented procedures, colour duplication as an end in itself should not be the goal in dyeing. It is often convenient to be able to duplicate a shade, but plants vary, even from one season to another. Careful records stating how much mordant has been used will enable the dyer to reproduce a colour if the dyestuff used is a fairly stable one (eg, onion, goldenrod).

ALL AMOUNTS GIVEN ARE FOR THE MORDANTING OF ONE POUND (453 g) OF SKEINED FIBRE OR LOOSE, PREPARED FLEECE. If mordanting less than this weight of fibre, decrease the amount of mordant proportionately. Mordants in powder or crystal form are thoroughly dissolved in boiling water after being measured, and liquids are diluted with at least an equal amount of water. Mordants in crystal form (eg, tin crystals) may be weighed rather than measured. If this is inconvenient, then the crystals can be first ground in a mortar and pestle and then measured.

In Table 2, volume is measured in millilitres (ml) and weight in grams (g). The new measuring spoons, available in sizes from 1 to 25 ml, are not the same size as teaspoons and tablespoons; however, in the smaller amounts (1 and 2 ml) they are so similar as to be almost interchangeable.

LITMUS TESTING OF MORDANT AND DYEBATH

The pH balance of a mordant or dyebath refers to its degree of acidity or alkalinity. This is important for the dyer to know, as excesses of either may impair the quality of the dyed fibre and inhibit the degree to which the fibre accepts the dye (Beth Parrott, 'The Chemistry of Dyeing' 56). On a scale from 1 to 14, 7 is neutral; numbers below 7 indicate acid conditions, while numbers above 7 indicate alkalinity. The strongest acid is

Table 2: Amount of mordant required*
(for 1 lb / 453 g fibre (for 4 oz / 114 g divide by 4)

	mordant	amount
Alum	aluminum ammonium sulphate	4 oz; 114 g
	aluminum potassium sulphate	4 Tbsp; 60 ml
Ammonia		2–3 tsp; 10–15 ml
Baking soda		2 Tbsp; 30 ml
Blue vitriol	copper sulphate	2–3 Tbsp; 30–45 ml
Chrome	potassium dichromate	2–3 tsp; 10–15 ml
Cream of tartar		4 tsp; 20 ml (more will not harm the fibre)
Glauber's salts		4 Tbsp; 60 ml (more will not harm the fibre)
Iron	ferrous sulphate	1–2 tsp; 5–10 ml
Salt		½ cup; 120 ml
Tin	stannous chloride	1–2 tsp; 5–10 ml scant, powder form 3.5–5 g scant, crystal form
Urine		2 qt; 2.5 l (more if urine is fresh)
Vinegar	acetic acid	1 cup; 240 ml (more will not harm the fibre)
Washing soda		2 tsp; 10 ml

* Approximate measures: ¼ tsp – 1 ml; ½ tsp – 2 ml; 1 tsp – 5 ml; 2 tsp – 10 ml; 1 Tbsp – 15 ml

Table 3: pH balance of a dyebath

acid bath	neutral bath	alkaline bath
pH 1–6	pH 7	pH 8–14
litmus turns red	litmus doesn't change	litmus turns blue
vinegar, cream of tartar	no mordant	ammonia
onion bath: gold	onion bath: yellow	onion bath: bronze-green
Umbilicaria: barn red	*Umbilicaria:* orchid	*Umbilicaria:* magenta-purple

therefore 1, and 14 the strongest alkali (see Table 3). By using a neutral metal dyepot (enamel, stainless steel, glass) and water that is as neutral as possible (see p 30), the dyer can control the acidity or alkalinity of a bath by the addition of certain mordants. Because wool is highly sensitive to alkalis, it must be processed at low temperatures when an alkaline substance is present.

Litmus paper, purchased from druggists or chemical suppliers, is used to test for the presence of acids or alkalis in the dyebath. If a neutral bath is desired (pH 7), the dyer first tests the dye liquor with litmus and takes a reading. If it is, say, mildly alkaline (pH 8–10), a small amount of vinegar or cream of tartar may be added to neutralize the bath. However, if the dyer wishes it to be even more strongly alkaline (pH 10–12) in order to produce a specific shade, then a small amount of ammonia could be added to increase the pH. A dyer wishing an acid bath tests the pH first, and if it is between 6 and 7, adds more vinegar (or another acid) to obtain a reading of 2 or 3.

As Table 3 indicates, the acidity or alkalinity of a bath to a great extent determines the colour. This is perhaps only of interest to the serious dyer, but even the beginner should be aware of the chemical changes that occur with the addition of certain mordants.

WATER

In a region notorious for its hard water, any discussion of plant dyeing must be prefixed with an explanation of what hard water is and the methods available to soften it. Water hardness is measured by the number of grains of calcium carbonate per gallon of water (gpg). Water containing more than 7 gpg is considered hard. In households where there is a commercial water softener, that water may be used for dyeing. Additions of vinegar or, even better, acetic acid, will soften hard water, as will the use of small amounts of laundry softening agents, particularly zeolite and Calgon (Elsie G. Davenport, *Your Yarn Dyeing* 49, 50). Softening agents such as Fleecy may be added to the final rinse after dyeing, but can discolour soft and delicate shades. Even fresh stream water may be hard, especially if it flows over mineral beds. Leechman recommends testing all water with litmus, and then adding enough acid to neutralize the bath (7 on the scale) (Douglas Leechman, *Vegetable Dyes from North American Plants* 19).

Collecting fresh rainwater provides the dyer with a reasonably uncontaminated source, unless there is air pollution. Gathering rainwater sounds more onerous than it is. Wooden barrels make excellent catchalls, although the tannic acid of the wooden barrel will affect colours by darkening them. It is not known whether or not the tannic acid leaches out after a period of time and more or less neutralizes the effect of the wooden collection barrel. Wooden and leather buckets were commonly used in the Atlantic region by homesteaders in the mid-1700s (Robert Cunningham and John C. Prince, *Tamped Clay and Saltmarsh Hay* 10), so presumably tannic acid from this source is not toxic, unlike its powdered chemical counterpart (*Journal of the Chicago Horticultural Society* 19).

Water collected from newly shingled buildings may be affected by the preservative now applied to new wood and asphalt shingles. In the case of the latter, this may appear as a greasy scum floating on top of the water in the collection barrel. After a few months it should be safe to use water from a newly shingled roof without any discolouration of the dyed fibres. Experiment first, and if the shades seem satisfactory, use the water for dyeing.

Plastic pails can be used to collect rainwater, as well as the dyepots themselves. However, use of dyepots as collectors does mean the dyer risks having enamel pots rust and corrode. In winter, snow can be melted for dyeing, provided it is clean and free of debris. City dyers will probably have to place mesh or screening over their collection barrels to keep out soot and other dirt. Bugs and leaves can easily be skimmed off the top.

The hardness of water and the condition of the pipes through which it runs contribute to the colours a dyer obtains. Although city water is generally considered to be hard as well as heavily chlorinated, urban dyers will not necessarily have unattractive colours. In fact, they will probably be brighter than those achieved by the country dyer using soft spring or rainwater. I live where there are many gypsum deposits, and my colours are greatly affected by this, resulting in a predominance of greens and rusts.

GENERAL DIRECTIONS FOR ALL METHODS OF MORDANTING

In the event that the following directions lead beginning dyers to believe the process is never-ending, it should be pointed out that today's dyeing techniques have been considerably streamlined. Weaver Patricia McClelland of Summerville, Nova Scotia, tells the following interesting story about a diligent dyer, as recounted to her by the late Ellis Roulston, who was chief of handcraft instruction, Department of Education, Halifax. It seems the woman in question, who was Scandinavian by birth, each July would travel to Peggy's Cove, located about thirty-five miles from Halifax, and there laboriously gather from the rocks the fresh droppings of seagulls who fed daily on the blueberry crop. Apparently a dyebath was made from this unique dyestuff, although no one knows whether the droppings served as the dyestuff or a mordant. (An unidentified Micmac woman told me of using the droppings of raccoon and porcupine, which feed on late summer berries, for a similar purpose.) Presumably they could be both. Most spinners and dyers are aware that that portion of a sheep's fleece which has been stained by deposits of urine and faeces takes on a much brighter colour when dyed than other, uncontaminated parts of the fleece.

For a description of the different methods of mordanting, turn to page 35. To prepare fibre, see Chapter 5. To mordant fibres other than wool, see pages 45–50.

1 Use plastic spoons for measuring mordants, and glass or plastic cups in which to dissolve them. NEVER USE DYEING EQUIPMENT FOR FOOD COLLECTION, STORAGE, OR PREPARATION. KEEP ALL EQUIPMENT OUT OF THE REACH OF CHILDREN AND PETS.
2 Dissolve powdered mordants thoroughly in boiling water, and dilute liquid mordants. Mordants in the form of crystals may be ground first using a mortar and pestle. Use the amount of mordant given on page 29. When in doubt, use less rather than more. Wear rubber gloves when handling chemicals.
3 Thoroughly wet the fibre to be mordanted. Handspun yarns, silk, cotton, wood, bone, and grasses should be soaked out overnight if possible. All fibres benefit from a long soak prior to mordanting or dyeing. This expands the fibres fully and ensures total and even penetration of the mordant or dye.
4 Add the dissolved mordant to several gallons of lukewarm water in the mordant pot. Then place the wetted fibre in the pot, making sure there is enough water to cover it completely. The fibre should be able to 'swim' freely in the bath. Never pour a mordant mixture directly onto the fibre in the pot. Always remove the fibre first (if it is already in the pot, as is the case when mordanting and dyeing simultaneously), add the mordant, stir well, and put the fibre back in.
5 Do not stir yarn about in the pot or push it down against the bottom where the pot is hottest. Hard stirring will cause the fibre to shrink and felt, and forcing it into constant contact with the hot part of the pot will result in streaky mordanting, and streaky dyeing. Gently lift and turn over the fibre in the pot, moving it around from time to time so that no one part of the skein is more exposed to the bottom of the pot than any other.
6 Start with lukewarm water in the mordant bath, and gradually raise the temperature to a simmer (200°F, 95°C) over an hour. Once a simmer is reached, time the bath for an additional 30 to 45 minutes. The mordanted fibre may be left in the bath overnight, which results in extra fastness and usually stronger, deeper colours. The mixture of mordant bath and fibre should be turned out into a plastic bucket and covered. This protects the mordant or dyepot from premature rusting.
7 After mordanting, the fibre is rinsed in successive baths of very hot, then cooler, water. The first rinse should be in water which is the same temperature as the mordant bath from which the fibre has just been removed. Avoid sudden temperature changes in the rinse water to minimize shrinkage (see shrinkage, p 50).
8 After rinsing, allow the fibre to drip from a plastic or stainless steel colander. Never wring wet fibre, or twist it to remove excess moisture. Gently squeeze the water out with the hands. Avoid letting mor-

danted or dyed fibre drip in the sink rack used for drying dishes, as minute traces of poisonous chemicals may remain on the utensil.

9 The usual procedure after mordanting is to use the yarn right away or dry it thoroughly for future use. Wet mordanted fibre can be stored for a few days in a plastic bag, but this is not recommended.

Please note If pre-mordanting with IRON: use a separate mordant and dyepot reserved for iron processing. If pre-mordanting with TIN: total processing time should be no longer than one hour, to minimize impairing the quality of the fibre. (Iron and tin are usually not pre-mordanted: see saddening and blooming, p 39.) If pre-mordanting with CHROME: work at dusk or with the blinds drawn. Keep the pot covered except when turning the yarn over. The light sensitivity of chrome is debatable (see p 25) and depends upon individual preference.

RINSING, DRYING, AND STORING MORDANTED YARN

Different authors prefer various methods for handling mordanted fibre. Seonaid Robertson (*Dyes from Plants* 27, 28, 30) rinses iron and tin after mordanting (although she prefers both of these mordants to be used as saddening or blooming agents, after dyeing). She does not rinse alum, chrome, or blue vitriol mordanted fires, however (ibid). Alma Lesch (*Vegetable Dyeing* 22–3) rinses after using all mordants, and this is what I recommend. Elsie G. Davenport (*Your Yarn Dyeing* 66–9) does not rinse after mordanting with alum, but does with chrome, iron, and tin. As with every other aspect of plant dyeing, techniques vary. Experiment and use that method which best suits you and your dyeing practices.

Several references stress the need to store chrome-mordanted fibres in the dark, but unless the yarn is dry it may mildew. Dyers who wish to treat chrome as if it were light-sensitive can dry their skeins by placing them in a dark garbage bag with an old towel. Place this in a warm spot, and occasionally pick the bag up and turn it over. When the yarn is almost dry, untie the bag, remove the towel, and allow the fibre to finish drying completely. Then store in a dark place or a garbage bag.

Wet skeins of mordanted (or dyed) fibre may be dried in any of the following ways: hang outdoors in the shade on a clothes-line or wooden rack; hang indoors over a wooden rack; leave to lie in a dry bathtub. Although most dyers do not recommend using a clothes-dryer to dry mordanted or dyed fibre, this is the method used by some people. I cannot suggest that it is safe, however, as there is too great a danger of shrinkage.

Some dyers weight down their skeins using a variety of methods, such as inserting a wooden rod through one end of the hanging skeins, or tying on a bottle filled with liquid. Handspun yarns may benefit from being stretched in this method, as they sometimes crimp after mordanting and dyeing. However, most commercially spun fibres do not. Whether weighted or unweighted, skeins should be turned frequently to prevent an excess of mordant (or dye) from dripping onto one part.

Pre-mordanted and thoroughly dried fibres may be stored for future dyeing. If not light-sensitive, the skeins should be placed in unclosed plastic bags. If keeping a large amount, moth balls may be added, enclosed in a separate container within the bag of fibre. This will allow the fumes to penetrate without the balls coming into contact with the fibre. Given the diversity of dyeing methods available to the dyer, it seems unimportant to store large amounts of previously mordanted fibres unless a special project or workshop is planned. In any event, the storage period should be limited to several months.

IDENTIFICATION OF PRE-MORDANTED FIBRES

Dyers who pre-mordant yarns and store them for future use require a means of identifying which yarn has been treated with what mordant. Some mordants impart a characteristic colour to a fibre: for example, yarn pre-mordanted with blue vitriol has a pale bluish-green tint. However, alum- and tin-mordanted yarn are almost indistinguishable, except by smell.

There are two convenient methods for identifying pre-mordanted fibres. The first involves using a specific fibre as a tie for each mordant: for instance, an alum-mordanted skein would have linen ties; a blue vitriol skein, cotton ties; a chrome skein, tan seine twine, and so on. Some dyers use linen for ties, with one tie designating alum, two blue vitriol, three chrome, and so on.

The second method is the knot identification system used by Miriam Rice (*Let's Try Mushrooms for Color* 23) and introduced to me by Dawn MacNutt. I have adapted it somewhat. Using five skeins of wool yarn and five mordants (alum, blue vitriol, chrome, iron, and tin), pre-mordant each skein with one mordant each. Tag each skein to designate the mordant used, and allow them to dry thoroughly. Then remove a twelve-inch length of yarn from each skein and tie a knot at the end of this piece to designate which skein it came from: one knot for alum mordant, two for blue vitriol, three for chrome, four for iron, and five for tin. Then make a small hank consisting of a twelve-inch length of yarn from each of the five skeins. Knot these together at the top. This then serves as both a sample hank to test dyestuffs and a 'colour chart.' The hank may be cooked with a small amount of dyestuff to see if that plant warrants further investigation. Because five mordants have been used, the hank of yarn will show the dyer the complete range of colours a plant will yield, and at a single glance. If you put one of these hanks in each dyebath, along with the skein to be dyed, you can immediately tell which mordant the skein was treated with by 'reading' the test hank. Some dyers prefer to tie the knots on the ties that they put around their skeins. Others use the more conventional methods of tagging skeins with garment tags, although these tend to get wet and blur as the skein drips dry.

MORDANTING: FOUR METHODS

There are four methods of mordanting fibres to be dyed with plants: pre-mordanting; simultaneous mordanting; the pot as mordant; and 'saddening' and 'blooming.' The latter are carried out after the dyeing has been completed, and as their names imply, either drab or brighten the colour already obtained from the dyebath. No one method is necessarily better than another, although each dyer soon develops a preference for certain techniques. Try all four, and then use that which is most suited to your facilities and the time available for dyeing. Although pre-mordanting is time-consuming, it is a popular method with such mordants as alum and chrome. Simultaneous mordanting lends itself to the workshop situation, while using the pot as a mordant is an appropriate technique for the school classroom. Saddening and blooming are often preferred by experienced dyers who wish to control the colour they get right up until the very end of the dyeing. Some dyers combine the methods: for example, yarn which is bloomed, or brightened, with the addition of tin at the end of the processing may also have been previously mordanted with, say, alum.

Pre-Mordanting

As the term implies, with this method the mordanting is done before the dyeing. This technique has the advantage of enabling the dyer to have on hand a number of mordanted skeins ready to be dyed at a moment's notice. Pre-mordanting results in even dyeing, and the yarns thus processed are usually quite fast to washing, an important factor if the dyed fibre is to be used to make garments or household articles. Some dyers pre-mordant all their yarns, and they claim that the colours obtained have a special richness. The one disadvantage with pre-mordanting is that it takes double the amount of time, as the fibre is processed twice:

once in the mordant bath, and once in the dyebath itself. Alum, blue vitriol, and chrome are most often pre-mordanted, while iron and tin rarely are. Iron and tin may be pre-mordanted when using the knot identification system explained on page 34. Otherwise, it is considered risky to process yarn twice with iron and tin as too high a heat can impair the fibre, as can too long a processing time.

Procedure for Pre-Mordanting

1 Dissolve the mordant in 4 cups (1 l) of boiling water in a glass measuring cup. Stir well, until it is completely dissolved. Add this mixture to 4 gallons (18 l) or more of warm water in the mordant pot. Stir well.
2 Enter 1 pound (453 g) of thoroughly wetted, skeined yarn. If pre-mordanted fleece, put it in the pot loose. Do not put skeined fibre and loose fleece in the same mordant or dye bath as they will tangle around each other.
3 Cover the mordant pot, and start to raise the temperature until the bath reaches a bare simmer (200°F, 95°C).
4 Lift the cover and occasionally move the yarn around to ensure even penetration of the mordant. Do not 'stir' the yarn as such. If pre-mordanting with chrome, an old plate can be put on top of the skein in the dyebath to keep it submerged. However, lift this off and turn the yarn often so the fibre does not scorch or streak.
5 Process the fibre for 1 to 1½ hours. Add more water, if required, so that the yarn swims freely in the bath. Keep the pot covered. Do not let the temperature of the bath exceed a simmer.
6 When the processing time is up, the fibre may be left to cool in the mordant mixture overnight or rinsed and dried.
7 Rinse in several baths of water, as described on page 32 (general mordanting procedures). Dry, or use immediately in a dyebath.
8 Tag to identify mordant used if the yarn is to be stored.

If pre-mordanting with chrome, the above procedure should be carried out in a darkened room or at dusk on an overcast day. As previously explained (pp 25, 33), the light sensitivity of chrome is debatable. Tests done using pre-mordanted chrome yarn in an onion bath, with one pot covered and the other uncovered, revealed that the covered bath produced a darker rust shade of a richer hue than the uncovered bath. However, both colours were attractive and satisfactory in every other way.

Simultaneous Mordanting

Simultaneous mordanting occurs when the mordanting and dyeing are carried out at the same time, in the same bath. The advantage of this method is that the yarn is only processed once. The total process often takes barely an hour to complete. This is an important consideration when using iron or tin as mordants. Simultaneous mordanting is well suited to the workshop and classroom situation as little time is lost, and the results can be seen almost immediately. When tin is added to a strong onion bath, for instance, the wet fibre takes on a brilliant colour almost the moment it is entered in the dyepot. This visual impact is an important feature of public demonstrations, where there is little opportunity to explain lengthy procedures.

Procedure for Simultaneous Mordanting

1 Cook out the dyestuff in water to cover. Strain off the plant and save the dye liquor, which now becomes the dyebath.
2 Dissolve the mordant selected in boiling water and add it to the dyebath. Stir well.
3 Enter the thoroughly wet fibre. (Yarn should be soaked out in water which is the same temperature as the dyebath.)
4 Cover the dyepot, and raise the temperature of the bath to 200°F or 95°C, for an hour.
5 Uncover the pot occasionally and move the fibre around. Check the colour, remembering that the dry colour will be lighter than the colour of the yarn when wet. If a different colour is desired, additional mordants may be added. Iron or tin added to the bath during the final fifteen minutes of processing will drab or brighten the resulting colour, whereas ammonia or baking soda will 'green' it.
6 When the desired colour has been obtained, lift the fibre out of the dyepot and squeeze the water from it. Then rinse in several baths of successively cooler water until all the excess colour appears to have been removed. (This bleeding out of the extra pigment is normal.)
7 If desired, more fibre may be added to the so-called 'exhaust bath,' provided there seems to be sufficient colour left in it. Additional mordants may also be added, but should be in amounts smaller than those normally recommended (see p 29). The cooking period for additional fibre will vary, depending upon the desired colour. Most exhaust baths are fairly pale, but some (goldenrod, onion, *Umbilicaria*) yield good colours after several 'dips.'
8 Dry the fibre. If storing it for future use, place the dry yarn in plastic bags closed with a twist tie.

Note The addition of extra water to the mordant or dyebath during processing, with any mordanting or dyeing technique, does NOT dilute the effect of the mordant or dyestuff. Adding yarn, however, will do just that. The amount of plant pigment in suspension in the water solution is determined by how much dyestuff was used and not by the volume of water in which it is processed. But using a generous amount of water does make for more even mordanting and dyeing.

Always decrease the amount of a mordant used if more than one is added to a bath. If yarn has been pre-mordanted with alum, and you then place it in a dyebath of onion with tin, use less tin than the regular amount. If adding blue vitriol and iron to a simultaneous bath of goldenrod, use the same amount of blue vitriol as recommended for the weight of the fibre, but only half as much iron.

The addition of iron to a simultaneous bath is sometimes not anticipated. The dyer may be using a regular pot, rather than the one kept just for iron mordanting and dyeing. If this is the case, pour the mixture into the separate iron pot before adding the iron mordant. If you have no pot for iron processing, the addition of a small amount of ammonia often produces a similar result, although the colour will be somewhat brighter. Substituting blue vitriol and salt for iron results in colours that appear iron-mordanted.

If using baking soda as a mordant in a simultaneous bath, do not add it until the final fifteen minutes of processing. Allow the dyebath to reduce so that the dyepot is no more than half full. Take the fibre out and add the soda; let the bubbling subside, and then re-enter the yarn. Do not cover the pot. Keep the temperature below 200°F or 95°C.

Yarns simultaneously mordanted and dyed with iron, tin, ammonia, or soda should be rinsed in water to which a small amount of non-detergent soap has been added (Ivory, powdered, or dishwashing liquid). Soap may or may not be used in rinsing all mordanted and dyed yarns. It is more a matter of personal preference than of technique.

Pot as Mordant

Aluminum, iron, copper, brass, and tin pots may be used as the actual mordant when dyeing is done with such a utensil. Using an aluminum pot affects the colours obtained much the way using alum as a mordant would. Tin pots brighten the yarn dyed in them, and the use of copper and brass pots results in lovely rusts and greens. This method of mordanting is well suited to the elementary school classroom, as there are no dangerous chemicals to be handled and no toxic fumes produced. It is, in one sense, costly, as the pot used cannot be used subsequently for food preparation. However, one dyepot can serve many dyers, and those who invest, as a group, in antique pots will never regret the expenditure.

Procedure

1 Prepare the dyebath, using a non-neutral metal pot, and strain off the cooked plant material.

2 Allow the bath to cool until it is lukewarm, or add ice cubes to it.
3 Enter the wet fibre to be dyed.
4 Cover, and allow the heat to reach 200°F or 95°C over the period of an hour. Uncover the pot occasionally and turn the yarn over.
5 Remove the fibre from the dyebath when the desired colour has been obtained.
6 Rinse the fibre in successive baths of water, the first of which is the same temperature as the dyebath.
7 If there appears to be any colour left in the bath, additional wet fibre may be added for an exhaust dip.

Note Additional mordanting may be carried out during the processing by adding dissolved mordants in the regular amounts. However, if more than one mordant is subsequently added, decrease the amounts.

As previously mentioned (the pot controversy, p 12), many dyers have considerable disdain for this method of mordanting and dyeing, which is, in fact, a type of simultaneous mordanting. But as long as dyers understand how reactive metal dyepots affect the dyeing process, there seems to be no valid reason for not using any kind of vessel whatsoever. Fall marigolds, cooked in a copper pot, yield a colour that every dyer should have the pleasure of seeing firsthand.

Saddening and Blooming

A dyebath is 'saddened' when iron is added at the end of the processing, just before the dyeing is done. If the mordant added at this time is tin, the bath is 'bloomed,' or brightened. Essentially, it is mordanting after dyeing, as the pigment has already been accepted by the fibre. What the dyer does, then, is to influence the value and hue of that colour by adding iron or tin. Saddening and blooming are processes that allow the dyer to use these mordants without worrying about destroying the quality of their yarn, provided the temperature of the dyebath remains BELOW 200°F or 95°C. The addition of a small amount of ammonia at the end of dyeing will 'green' a strong yellow bath such as onion or marigold. The resulting colour is brighter than an iron drab, but not as drastic a change as if the tin had been added.

Procedure

1 Prepare the dyebath and enter the wet fibre.
2 Raise the temperature of the bath to 200°F or 95°C and process for forty-five minutes.
3 Dissolve the selected mordant in boiling water, remove the fibre, and add the mordant mixture. Stir well, and re-enter the fibre.

4 Process the bath for an additional five to fifteen minutes or turn off the heat completely. Longer cooking will impair the quality of the fibre, as will heat higher than 200°F or 95°C, and often excellent results are obtained without any heat.
5 Remove the fibre and rinse in several baths, to which soap has been added. A final rinse of water and vinegar (1 Tbsp, 15 ml to one gallon or 4.5 l of water) will neutralize the alkalinity of the iron, tin, or ammonia without altering the colour obtained.

Note Blue vitriol and iron, added to a dyebath of fresh lettuce, produced a good green after the bath was removed from all heat. The mordants were dissolved and added off the heat, and the fibre remained in the bath for several hours.

DYEPLANT AS MORDANT

As previously mentioned (p 27), some dyeplants are used in the pot as a mordant. This is true with alder and with sumac leaves. Barks, roots, nuts, and tree galls, all of which contain tannin, may be used as a mordant by adding the material to the actual dyestuff (ie, lichen) at the time it is being cooked out. However, this would require a larger than usual pot. Some dyers first cook out the dyestuff, strain it off, then add the dyeplant being used as a mordant, along with the fibre to be dyed. The tannin material may also be cooked out first, strained off, and then the liquid added to the strained off dye liquor. Tannic acid darkens tans and browns and is reported to darken fibre dyed with it, after prolonged exposure to light. However, a subsequent dip in blue vitriol, iron, or chrome supposedly will prevent this occurrence (Rita J. Adrosko, *Natural Dyes and Home Dyeing* 71).

5

Natural fibres and their preparation for dyeing

WHAT CAN YOU DYE?

Almost all plant and animal fibres can be dyed. Some fibres, notably wool and silk, dye more successfully and easily than others. Acrylics and other synthetics take a dye, as will such unlikely substances as human hair and bones. Various North American Indian tribes dyed their skin and hair, or, more precisely, stained it by rubbing on pigment with their hands. A colour made from hemlock and mud was used by the Chilkat for their famous blankets (Douglas Leechman, 'Aboriginal Dyes in Canada'). Modest pine furniture was stained with berry juices, and many of the oldtime paints contained blood.

Contemporary dyers use many fibres for dyeing, including polar bear fur (see Judy McGrath, 'The Dye Workshop'), feathers, and dog hair. Dog hair is spun into a yarn along with wool, which has a longer staple. Grasses, shells, leather, and wood all take a dye. The umbilicate lichens are even said to dye marble (Eileen Bolton, *Lichens for Vegetable Dyeing* 44). Today few native basket-weavers make plant dyes (see p 27), but in *Halifax, Warden of the North* Thomas Raddall describes 'Micmac baskets of maple splints, dyed in various colours' being sold in outdoor Halifax markets at the turn of this century. One young native dyer has experimented with porcupine quills she collects from carcasses found along roadsides (Vivian Grey, Halifax).

WOOL, THE DYER'S FAVOURITE

Pure wool, whether unspun fleece or yarn, is the most widely used fibre for plant dyeing. There are several reasons for its popularity. Always available, wool is relatively inexpensive, easy to handle, and diverse in its uses. Wool takes a dye well, and the colours produced are usually fast to light fading and repeated washing. Because of its washability, wool lends itself to the making of garments and household articles that receive hard wear. Whether knitted, woven, braided, macraméd, crocheted, or hooked, wool is unequalled in appearance, durability, and ease of care.

SCOURING WOOL

The term 'scour' describes the process of removing the natural oil and grease from wool. If left in, these oils would prevent the dye from penetrating the fibre and adhering properly to it. Although scouring often implies scrubbing, this is not the case. The fibre is, instead, carefully washed in warm soapy water, and never rubbed, or wrung out. Often successive baths are required, especially when the fleece is very dirty (see p 43 for processing commercially spun fibres). Using a non-detergent soap (Ivory) results in the scoured fibre retaining some of its oil, while using a detergent will remove all traces of the natural grease from fleece and handspun yarn. A yarn scoured with a detergent will dye a lighter colour (see below, bleaching). Some references (eg, *Journal of the Chicago Horticultural Society* 12) state that wool must be scoured with an alkaline substance (detergent, washing soda) in order to prepare the fibre to accept the acidity of most dyeplants. But alkalis may be harmful to wool, and I have found that unscoured, clean fleece and handspun yarn dyes extremely well. Therefore, dyers who use fleece or handspun yarn for dyeing are advised first to experiment using a non-detergent soap to wash the fibre prior to mordanting and dyeing. If this gives satisfactory results and the fibre dyes well, then there is no need to use a detergent. Some of the handspun yarns photographed for this book were soaked overnight prior to dyeing, with no soap used at all.

BLEACHING WOOL

It is not essential to bleach wool prior to dyeing unless dyers using handspun fibres wish to start with a very 'white' yarn. 'White' yarn is snow white, or bleached. 'Unbleached' or 'natural' white yarn retains its original yellow-white or grey-white colour. Household chlorine bleach or ammonia can be used. Allow 1 tsp (5 ml) of chlorine to each gallon (4.5 l) of water. Let the pre-wetted fibre soak in this rinse for half an hour. Turn the fibre over occasionally, and rinse afterwards in clear water. One recipe for a hydrogen peroxide bleach for wool suggests that bleaching is necessary to obtain light colours (Mary E. Black, *New Key to Weaving* 506). While it is a fact that 'snow' white yarn gives the lightest colours clarity, satisfactory light shades can be obtained using unbleached or natural white yarn even if it has not been washed first with a detergent or soap.

PROCEDURE FOR COMMERCIALLY SPUN WOOL YARN

Although one author, Ida Grae, routinely neutralizes all commercial yarn she uses for dyeing, most dyers find it sufficient to soak out commercial yarns in a warm water rinse (*Nature's Colors: Dyes from Plants* 201). Soap or detergent may or may not be added. If even after careful dyeing, the yarn is badly streaked, it may well be, as Grae suggests, that factories are inconsistent when processing their yarns. If so, try the water and baking soda bath she describes (2 Tbsp, 30 ml soda, 1 gallon, 4.5 l water).

Factory-spun wool fibres may also have a 'fugitive' colour, which appears as a slight pink or greenish cast on the yarn. This must be removed using soap. Sizing agents which aid in spinning the fibres may also account for commercial yarns dyeing unevenly. If such sizing is known to be present, the yarns should be soaked prior to mordanting and dyeing in a detergent and water bath. When a detergent is used, it must be carefully rinsed out in clear water before the yarn is processed further.

PROCEDURE FOR HANDSPUN YARN

(See scouring, p 42.) Owing to its greater elasticity and varying twist (sometimes looser, sometimes tighter), handspun wool yarn must be carefully handled before mordanting or dyeing to ensure that it does not shrink or lose its attractive character. If spun 'in the grease,' that is from unwashed or unscoured fleece, it then should be washed several times and left in a final clear-water rinse overnight. Handspun yarn is more likely to shrink than its commercially spun counterpart, so care must be taken to avoid wringing it out or exposing the fibre to drastic changes in water temperature during processing. Tie handspun skeins more loosely than commercial skeins, in at least eight places if the skein weighs more than 8 oz or 228 g. When drying handspun skeins, they may be weighted (see p 34) just enough to prevent them from crimping. However, too heavy a weight will remove elasticity and may cause the yarns to break later.

Handspun fibres are variously described as dyeing 'better' than commercial skeins, 'worse' than factory yarns, or simply 'different' from store fibres. But 'better' and 'worse' reflect individual preference, and are not qualitative statements of fact. A more reasonable approach is that of Jack Kramer, who writes: 'Many craftsmen prefer to spin their own yarn, but homespun yarn does not take color as readily as factory yarn; the dye color is not as bright' (*Natural Dyes: Plants and Processes* 28). There is no reason to disbelieve this statement. Indeed, my personal experience led me to the same conclusion. However, after an intensive summer dyeing workshop that I conducted for the Nova Scotia College of Art and Design in Halifax in August 1975, I changed my mind. Students who dyed handspun wool yarns (mostly from Leicester fleece) consistently obtained brighter shades when their skeins were in the same mordant and dyebaths as commercial yarns. Suffice it to say that handspun yarns may or may not dye differently from factory yarns depending upon the quality and type of fleece from which the yarn is spun, the processing of the fleece prior to spinning, the spinning itself, the mordanting, and the nature of the dyebath and the chemicals used. As suggested at the beginning of this book, dyeing is an individual skill, and the wise dyer offers no guarantees.

PROCEDURE FOR CLOTH

Dyeing cloth is tedious because most dyers do not have a large enough container to handle more than a small amount of fabric. Cloth must be able to swim freely in the mordant or dyebath in order to achieve even, thorough results. However, many hookers and quilters are able to dye one yard of wool cloth, either in a single piece, or cut into smaller pieces or strips, using a 10-gallon (45.5 l) pot. Commercially woven wool cloth (presumably white or a light colour) must first be thoroughly soaked out in warm soapy water. Take care to move the fabric around, but do not stir or agitate it unnecessarily. Use as much water as the pot will hold. Handspun wool cloth may shrink more than commercially woven cloth, depending upon the weave (see shrinkage, p 50). Cloth of handspun yarns should be soaked out overnight prior to mordanting or dyeing, with soap added to the water if the fibre was spun in the grease. When removing commercial or handspun cloth from a bath, let it drip out in a colander in the sink. Avoid squeezing the wet cloth, as this may cause it to felt. Commercial wool cloth may be dried by hanging it over a wooden rack after gently pulling it into shape. Handspun cloth should be blocked on a towel-covered board, or dried on a flat mesh sweater rack. It is essential to use a large pot when dyeing cloth, as too small a utensil will result in the cloth being crowded and mordanting or dyeing unevenly.

PREPARATION OF WOOL FOR DYEING

Note that the preparation of wool fibres for dyeing is the same as the preparation of these fibres for pre-mordanting.

1 Raw fleece should be thoroughly sorted, removing all debris (Elsie Davenport, *Your Handspinning* 32–8). It may or may not be carded,

although some teasing or fulling of the fibre with the fingers is recommended for maximum mordant and dye penetration. Wool dyed as fleece, before it is spun, is termed 'dyed-in-the-wool' (Black, *New Key to Weaving* 506). Soak out yarn in skeins; soak out cloth as described above.

2 To skein yarn, see pp 51–5. Do not place loose fleece in the mordant or dye bath with skeined yarn as the two will intertwine.
3 Bleach handspun wool if desired (see p 42).
4 Thoroughly dissolve non-detergent soap or soap flakes in lukewarm water, working up a good lather. (Use a detergent soap for oiled or handspun yarns when bright colours are desired from the dyeing.)
5 Move fibre around in the soap water. If the fibre is particularly dirty (as may be the case with raw fleece), several washings may be required. Some dyers use a plunger when processing fleece.
6 Rinse thoroughly in clear water the same temperature as the soap and water bath. Several rinses may be required to remove all traces of soap.
7 The fibre may be dyed immediately or left to drip out in a colander. Hang it up to dry or store damp for subsequent dyeing. (Damp wool can be kept for up to a week in a plastic bag, if the bag is turned over daily and kept out of extreme cold and heat.)

COTTON AND LINEN

Because cotton and linen do not have the same natural affinity for dyes as wool, they must be prepared differently for mordanting and dyeing. This is done by boiling them first in a bath to which a small amount of washing soda has been added (1 lb, 453 g fibre, water to cover, 1 tsp, 5 ml washing soda). This bath is processed for an hour. The cotton or linen is then removed and rinsed in clear water of a similar temperature. Unlike wool, cotton and linen are mordanted using what is described as the alum-tannin method (Seonaid Robertson, *Dyes from Plants* 88).

Most cotton and linen purchased for dyeing is either bleached white, or a natural shade such as ivory or soft beige. Some dyers bleach their own cotton and linen with household chlorine bleach (ibid). The yarn is then exposed to the air and resoaked and re-aired. A similar old method was 'bleaching cotton by steeping in lye and laying out on the grass' (*Eighty Years' Progress, 1781–1861, in the United States and Canada* 274). Bleaching bed linens in this manner was a common practice in rural areas of eastern Canada until the clothes-line became a status symbol.

Cotton and linen household articles, clothing, and cloth may be dyed. The success of the project depends upon the known fibre content of the article to be dyed. For instance, small handwoven linen articles dye well, but commercially woven linens may have been treated with a sizing at the factory. Such articles should be washed in warm soapy water containing washing soda or ammonia to remove these substances. Cotton T-shirts dye well, as do sewing notions such as eyelets and trim. Commercial cotton sheeting (with no synthetic content) dyes well, and is often used by quilters and doll-makers. Both cotton and linen respond best to strong dyestuffs – onion skins, hemlock branches, and marigolds.

ALUM-TANNIN MORDANT FOR COTTON AND LINEN

For the tannin, either tannic acid or fresh sumac leaves may be used (see sumac, p 209).

1 Soak out cotton or linen fibre for several hours.
2 For 1 lb (453 g) of fibre, dissolve half the regular amount of alum (as on p 29) in boiling water. Add this to two gallons (8 l) of water and stir well. If using sumac, read (3) below.
3 Dissolve 1 Tbsp (15 ml) of tannic acid in boiling water and add that to the bath. If using fresh sumac leaves, proceed as follows: Cook out 1 lb (453 g) of leaves in water to cover for one hour. Strain off the leaves and reserve the liquor: this contains the tannin. Add the dissolved alum to the tannin liquor in the mordant pot. Then add additional water so that the pot is two-thirds full before the fibre is entered.
4 Enter the wet fibre and raise the heat so the mordant bath reaches 212°F or 100°C in half an hour. Maintain this temperature for an additional fifteen minutes, and then remove the pot from the stove. At this point, the fibre may be left to cool in the alum-tannin bath overnight if desired.
5 It will be noticed after the processing that the fibre will already have achieved some coloration. This is normal. Sumac leaves contain a dye as well as the tannin mordant, and the fibre will have picked this up.
6 Rinse the mordanted fibre and dry it, or store damp for future use.

PROCEDURE FOR SILK

Silk, like cotton and linen, will dye but has less affinity for the dyestuffs than wool. It cannot be processed at a high temperature, and once processed and dried retains a characteristic odour that some people find disagreeable. This smell can be minimized by a final rinse, after dyeing, in Fleecy. However, use only a small amount, as too much Fleecy will discolour soft shades, especially yellows. Raw silk should be scoured prior to mordanting and dyeing, to remove the sericin, a natural gum coating (*Journal of the Chicago Horticultural Society* 14). However, as sericin is only dissolved by strong acids or weak alkaline solutions, this process is tedious because both substances may actually harm the silk itself. Therefore dyers wishing to purchase silk for dyeing should buy types which are ready to use. Such yarns are usually bleached white, or off-white, and suitable for a warp (ie, smooth, strong).

MORDANTING SILK

Note that iron and tin are not recommended as suitable mordants for silk, although Robertson (*Dyes from Plants* 95) suggests using small amounts of tin and the *Journal of the Chicago Horticultural Society* (p 31) uses iron for silk with oxalic acid. If you do use them, do so with care.

1 Thoroughly soak out the silk fibre to be mordanted and dyed. Then wash it in warm water (150°F, 66°C) and a detergent soap. Rinse in clear water.
2 If mordanting with alum, blue vitriol, or chrome use more than the amount recommended on page 29 for 1 lb (453 g) of fibre. Dissolve this in boiling water, and add it to a mordant pot two-thirds full of warm water.
3 Enter the wet silk, and raise the temperature so that the bath reaches 175°F or 80°C. (NOTE: this is below a simmer.) Maintain this temperature for two hours.
4 Turn the fibre over frequently, to make certain all parts of it are fully exposed to the mordants.
5 Rinse the fibre after processing. It may be dyed immediately, or dried for future use. (The disagreeable smell of wet silk seems to be made worse if the fibre is stored damp prior to dyeing several days later.)

Silk takes well to strong dyestuffs, such as onion skins and *Umbilicaria*. Silk dyed in an *Umbilicaria* bath, with alum, gives a soft rose; the exhaust dip yields an extremely attractive dull pink. With a chrome mordant, in an onion bath, a slubby off-white silk dyed a good lemon-yellow. In all cases, the fibre was left in the dyepot overnight (after first pouring it into a plastic basin). Pre-mordanted silk fibres give stronger shades than those mordanted simultaneously.

FUR AND HAIR

Animal fur and human hair may be successfully dyed. The problem here is how to contain the fibre during processing. Some dyers solve this by spinning the fur or hair first and then dyeing it. Wet fur or hair that is short is easily matted, so it must be placed in a bag made of muslin, cheesecloth, or cotton. Sew the bag so that it is large enough to contain the fur or hair without the fibre being crowded. If using muslin or cheesecloth, the mordant and dye will penetrate and affix itself to this as well as the hair or fur. However, if a synthetic cloth bag is used (or permapressed cotton), little dye will be wasted on the bag itself. Soak out fur or hair overnight in water to which a non-detergent soap has been added. A plunger may be used to slosh the water through the bag. Rinse the fibre and bag in clear water the following day. It may then be placed in the mordant or dyebath, and processed as for silk (ie, at a lower temperature than for wool). The fibre can be left in the bath overnight or for several days to intensify the colour obtained. White or light-coloured fur or hair dyes best, although some grey poodle clippings take on lovely shades after being dyed with onion or marigolds. Use strong dyestuffs when dyeing fur or hair and avoid mordants such as iron and tin.

PORCUPINE QUILLS

As mentioned earlier (p 41) porcupine quills are not commonly dyed now. Vivian Grey told me her best results were obtained when she added sugar to the dyebath to intensify the coloration. This is reminiscent of Ida Grae's blackberry dye mixture containing what a blackberry pie has as ingredients: namely the fruit, sugar, and flour (Grae 46). In other words, the sugar helps affix the pigment. Porcupine quills absorb dye slowly and benefit from a prior soaking in water and washing or baking soda. This alkaline bath helps soften the outer layer of the quill and prepare it to receive the pigment. Most dyers who wish to use quills will have to resort to Ms Grey's method. She removed the quills from dead animals along the roadside, which, as she explained, was hardly pleasant.

BONE, HORN, AND SHELLS

To dye bone, horn, and shells, first boil them out completely to remove all marrow, tissue, and attached debris. Fragile shells may break, so cushion them on a bed of fine sea sand or aquarium gravel. Then soak the cleaned bone (or horn, or shells) in a solution of 1 gallon (4.5 l) of water and 1 cup (240 ml) of household chlorine bleach. Remove after several hours, and if the material to be dyed is still not quite white, repeat the bleaching process using a fresh mixture. Drying the bone in the sun also helps to lighten the base colour. Prepare a very concentrated dyebath, using twice as much dyestuff as is required to dye wool. For example, to dye bones weighing 4 oz (114 g) you would require 8 oz (228 g) of dyestuff. Onion skins are appropriate for dyeing bone, as are all strong yellow flowers such as marigold and tansy. Use the same mordants as for wool, but double the amount. The bone may be boiled in the dyebath until such time as the desired colour has been obtained. Cool the bone in the bath. There is no need to rinse the material dyed after dyeing.

Bone, horn, and shells are useful to the craftsperson who makes garments or decorative hangings where found objects may be successfully incorporated into the design. Dyed bone buttons are an unusual touch when added to a handmade sweater knitted from plant-dyed yarns.

GRASSES AND FLOWERS

Dried grasses and flowers may be dyed, usually without heat. The soaked out material is simply placed in the prepared dyebath, to which a mordant has already been added. The grass or flowers can be held down in the bath with an old plate or a plastic colander filled with stones. Remove the weight after half an hour and turn the material over. Repeat this process half-hourly until the desired colour has been obtained. Dyeing grass and flowers in a hot bath provides brighter and more long-lasting colours, but the hot liquid softens the stems, which may be undesirable, depending upon how the material is to be used. Long grasses can be formed into a circle in the dyepot and weeds, flowers, or seed pods formed into bunches and tied with strings. Place them stem end up, unless the blooms are extremely fragile. Dry by hanging them upside down by the strings, and shake occasionally to prevent the pigment from building up in a single spot. Use double the amount of mordant suggested on page 29. Chrome, blue vitriol, iron, and tin produce interesting results, while alum changes the colour very little.

DYEING SISAL AND JUTE

Sisal and jute both will take a plant dye if they are carefully prepared beforehand. Sisal in the form of binder twine is treated with an oil which retards its decomposition in the hay field. This oil must be removed before mordanting and dyeing, by processing the sisal in a hot water bath to which a small amount of detergent soap has been added. Heat the bath to a boil, and then remove the pot from the stove. Let the sisal sit in the bath until it has cooled. Then rinse in clear water and proceed with mordanting and dyeing. Use double the amount of mordant recommended on page 29, for both sisal and jute. Jute may be soaked out prior to mordanting and dyeing in warm water to which a non-detergent soap has been added. Then rinse the fibre in clear water. Sisal and jute are easiest to process when they have been made into a skein. (Binder twine is used for baling hay, although in some areas it is now so costly a synthetic substitute is used. Look for sisal at country hardware stores and farm suppliers. Feed dealers are also a good source.)

DYEING SYNTHETICS

It may seem anachronistic to discuss dyeing of synthetic fibres with plant dyes, but for some would-be dyers today, synthetic yarns may be all that is immediately available. Teachers may have acrylics on hand, but no wool. Rather than discourage experimentation with dyeing, it is better to proceed with what you have in order to learn the techniques and in the meantime send for wool (see suppliers, p 228).

Most acrylic yarns will take a dye, and some dye brighter colours than wool. Acrylic ties on wool skeins dye brighter and darker colours with mordants such as chrome and tin, in onion and other strong baths. Few nylons dye well, and orlon may or may not, depending upon the type of yarn. Smooth yarns tend to dye best, especially if they are bleached white. Some rayons take a dye; again, it depends upon the yarn itself. There is no hard and fast rule. Try them all. Most synthetics cannot be harmed by processing at a high temperature, although some may not respond well to strong mordants such as iron and tin.

SHRINKAGE AND FELTING

There is a popular misconception that hot water alone causes wool to shrink, but this is not the case. Agitation is as important a factor in shrinkage, and a combination of both agitation and a high temperature will surely result in shrinkage.

All the precautions listed for soaking, rinsing, washing, scouring, mordanting, and dyeing wool are necessary to minimize shrinkage. A little will always occur, especially when processing woven articles or cloth, but the amount of shrinkage in a skein of yarn is minute if it is handled with reasonable care. Several processed (mordanted, dyed, dried) skeins were measured: one lost one yard (.914 m) in an 80-yd (72-m) skein, and the others less. The wool was a 2-ply medium weight. Shrinkage is only a problem to the dyer handling large amounts of yarn for a single project. If there is any worry that you may be several yards short in a particular colour, then dye a little more. An additional 2 oz (57 g) per 5 lbs (2 kg) of yarn should suffice.

Knitters are well aware that garments shrink when washed after they are knit, and allowances are made for this fact when working with pure wool yarns. Weavers who produce yardage intentionally shrink their cloth after it has been woven, by a process called 'fulling' or 'felting' (O.A. Beriau, *Home Weaving* 134). This involves soaping the wet cloth and agitating it with the hands, feet, or a paddle until the material has shrunk. Felting adds density to a cloth and prolongs its wearing ability.

FACTORS CAUSING UNWANTED SHRINKAGE

– excessive heat (above 200°F or 95°C)
– excessive agitation or stirring
– wringing or twisting wet fibres
– plunging fibre from a hot bath to a cold, or vice versa
– drying fibres in a clothes-dryer (see p 33)
– pressing wet fibre or cloth with a hot iron

MAKING A SKEIN

To dye yarn successfully, it first must be made into a skein or a continuous hank, which makes the fibre much easier to handle during processing. Most fibres intended specifically for weaving are already skeined and need only to be retied before mordanting and dyeing (see p 54). But many yarns, especially those used for knitting, come in balls. These should be rewound into skeins of varying sizes and weights before dyeing. It is convenient to think of yarn in amounts of 4 oz, 8 oz, and 1 lb. In metric measurement, these translate into:

1 g	=	.035 oz
28.5 g	=	1 oz
114 g	=	4 oz
226.5 g	=	8 oz
453 g	=	1 lb
1 kilo	=	2.205 lbs

The circumference of a skein may vary greatly, depending upon whether the yarn is commercially spun and skeined or handspun. Some skeins made up for workshops are as small as twelve inches (30.4 cm) in circumference, while 54 inches (1.37 m) is a standard size for commercial yarn factories. Skeins with a circumference of 80 inches (2 m) and more were commonly wound on niddy noddies in the last century (Dorothy K. and Harold B. Burnham, *'Keep Me Warm One Night'* 38).

Much of the antique spinning and weaving equipment presently contained in the Royal Ontaio Museum collection is from Nova Scotia. Pictured in the Burnham book are clock reels for skeining yarn which have counting devices that click when 100 yards are wound. Excellent examples of antique equipment, including skeining devices, are to be seen in Cape Breton at the Heritage Museum, along with an extensive collection of early weaving artifacts. (Located at Northeast Margaree, Heritage Museum is owned and operated by Florence MacKley of Sydney, assisted by her daughter, Marjorie Hart.)

Although reels, skein winders, and niddy noddies make the 'best' skeins (ie, the most uniform in tension), other equipment may be pressed into service depending upon the dyer's circumstances. Too often, interest in a new craft skill diminishes because the proper equipment is not available. What matters is not how you wind a skein, but that you do it. Learn to improvise.

REEL WINDING

Several types of equipment may be used to wind a skein, but a reel or a skein winder is the most efficient device. Elsie G. Davenport calls this a 'wrap reel.' Although expensive, a reel is a good investment for the serious craftsperson who skeins many pounds of yarn a year. Knitters find reels handy, as skeins may be wound into balls using such a device. Most reels have metal 'arms' which can be shortened or lengthened to accommodate skeins of varying circumference. However, antique reels and winders usually do not have this adjustable feature. Dyers using such equipment should wind their skeins loosely so they can be easily removed from the arms of the reel without undue stretching or tugging. Old reels seen at auctions and second-hand stores often have one or more missing arms, but replacements are easily made by woodworkers with a lathe. However, since an unusable antique reel may cost as much as a new, adjustable one, dyers should think carefully before paying for the charm of an old reel. If an antique winder is expensive, be able to verify that it is in some way significant (ie, has a counter; was made by a famous chair-maker, and so on) or ask for a statement of authenticity. Old spinning and weaving equipment is no longer given away or relegated to the trash heap.

To wind a skein using a reel, secure the end of a ball of yarn to one of the reel's arms. This end may simply be wrapped around several times or tied. Then turn the reel clockwise until sufficient yarn has been skeined, or the ball used up. Do not let the yarn pile up in the centre of the skein as it is wound, or at the edges. Feed it evenly, keeping the fibre under an even tension with your fingers. When the desired amount has been wound, tie the end of the skein through the skein itself in a figure eight, or else tie it back onto the end of yarn you began the skeining with. Make six or eight ties through the skein, keeping them loose enough to allow the mordant and dye to penetrate and yet not so loose as to allow the fibre to tangle.

TYING SKEINS

When mordanting and dyeing wool, most skeins are prepared for the dyepot by tying them with ties of cotton string, acrylic yarn, or perhaps

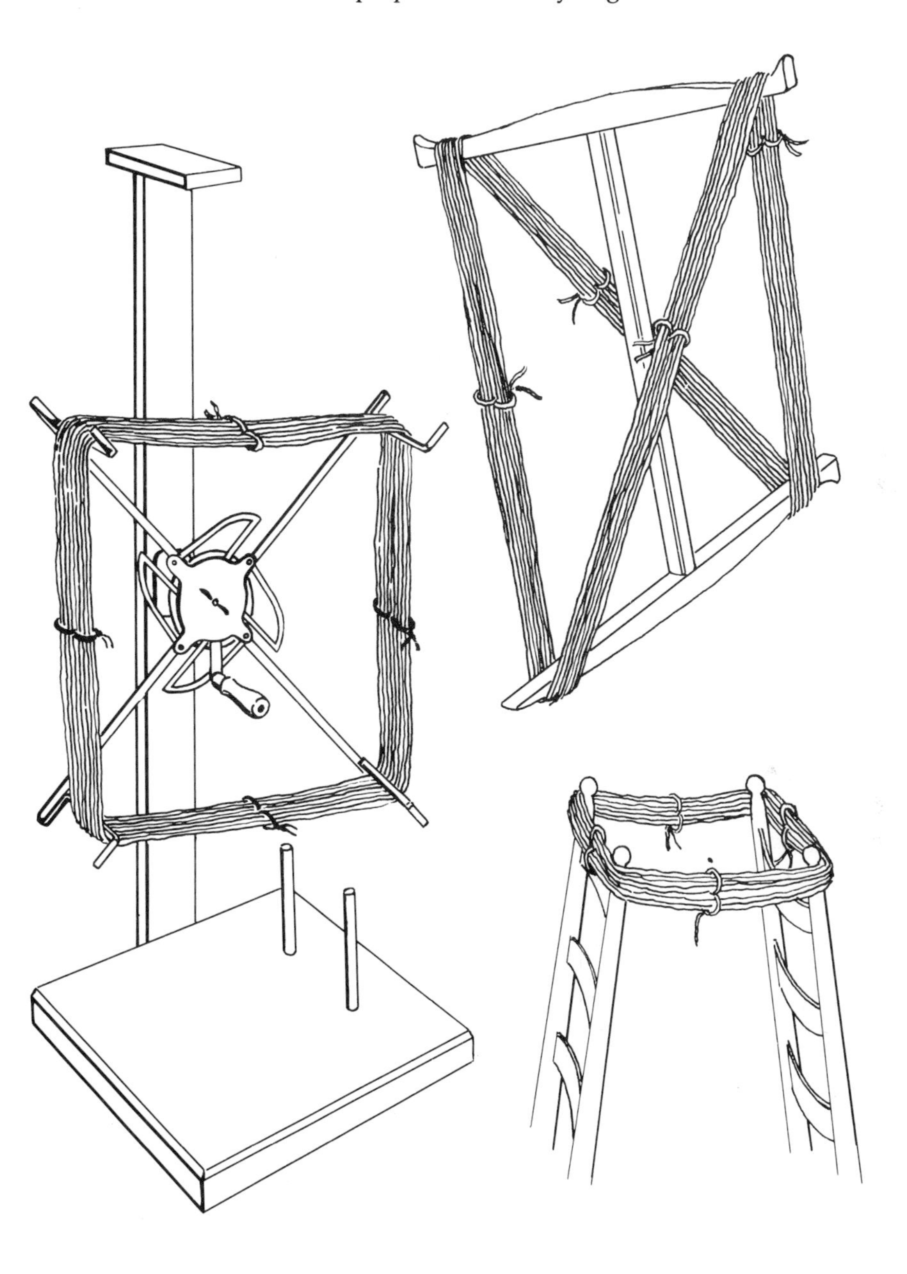

Skein winding: three methods

REEL NIDDY NODDY CHAIR-TO-CHAIR

linen. Ties which do not stretch are preferable; hence these fibres are more suitable for ties than wool. Ties which are too tight will result in streaked yarn. However, if this tie-dye effect is desired, then the ties are put on very tightly. Handspun wool yarn to be dyed should be tied in six to eight places with a stout cotton or linen yarn. Some dyers use specific ties to denote which mordant a skein has been processed in: cotton ties might indicate alum; linen ties, chrome; acrylic, iron, and so forth. (See p 34 for other identification systems.)

To tie a skein, cut six to eight pieces of fibre, about 6 inches (15 cm) long. Divide the skein in half with your fingers, and insert a tie through the skein and bring it back to tie onto itself, forming a figure eight. Some dyers use one end of the skein itself for the ties, so that they are all formed from the same piece of yarn and thus joined. This is fine, so long as the dyer is sufficiently experienced to realize that pulling on one tie may result in the next tie tightening and thereby prevent the mordant and dye from reaching that part of the skein. Ties which are too loose are far better than those which are too tight. You should be able to pass three fingers through a tie. Remember that fibre expands when wet, and a snug tie on dry yarn may prove to be a tight one on a wet skein.

NIDDY NODDY

A niddy noddy is a traditional skein-winding device that is simple and efficient to use, as well as fun. It consists of two bars of wood set at right angles to each other, one at each end of a piece of wood about eighteen inches long. (See illustration.) A skein is wound from the botton bar to one side of the top, back again to the bottom, and then up again to the other side of the top bar. Niddy noddies are available from some suppliers (see suppliers, p 228) as an alternative to a reel.

SWIFTS

Although it is often thought that swifts are not suitable as skein-winders (eg, Burnham and Burnham 38), Beriau (p 36) describes a vertical squirrel cage swift as a 'skeinwinder now widely used.' The top cage can be moved up and down, so the swift is adjustable. The Burnhams are probably referring to the more familiar 'umbrella' swift, which expands outwards rather like the ribs of an umbrella. This swift can be screwed onto a table top. Some of the modern Scandinavian ones clamp onto the loom. Although a swift is used for the exact opposite of skeining, that is, unwinding a skein into a ball of yarn, there appears to be no reason not to use it for the other purpose as well. Lacking a reel, an umbrella swift is probably as convenient a skein-winder as any other device. Many varieties of swifts are pictured in the Burnhams' book and interesting antique examples may be seen at the Heritage Museum in Cape Breton.

CHAIR-TO-CHAIR

Craftspeople are usually inventive, and rather than put aside the urge to learn something new, they improvise equipment. The chair-to-chair method of winding a skein is such an improvisation. To wind a skein, position two high-backed chairs as shown in the illustration on page 53. Two chairs 11 inches (28 cm) apart will make a skein with a circumference of 60 inches (1.5 m). Secure one end of a ball of yarn by tying it to one of the chair posts. Then wind the yarn around the four chair knobs until the skein is the desired size. Untie the end you started winding the skein with and retie that to the finishing end, in a figure eight, making the tie go through the skein itself. Then tie in five to seven more places. Do not hang up skeins after they have been wound; lie them flat. Any unevenness in winding makes the skein more likely to tangle in the mordant or dye bath. Skeins hung by their ties may pull and become quite uneven.

CLAMP METHOD

Another improvisation that works for skein-winding is to use C clamps. This idea is based on Mary Black and Bessie Murray's idea of using C clamps to wind a warp (*You Can Weave* 8). Position the clamps along the edge of a counter or table, making certain that the clamps are free of grease if they have been borrowed from someone's workshop. Wind a skein of whatever circumference is desirable or convenient.

UNTYING COMMERCIAL SKEINS

Commercially wound skeins should be untied and redone before mordanting and dyeing. This is because the single tie made at the factory is far too tight to allow the mordant and dye to penetrate the skein. It is easier to undo this tie if the skein can be stretched out on a reel or a swift. This also facilitates placing of additional ties required.

6

Dyeing procedures

AN APPROACH TO DYEING

Given the diversity of dyeing methodology, no dye book can possibly cover each approach to the satisfaction of each reader. Therefore, what is presented here is an amalgam of procedures from which each dyer or would-be dyer can extract the information best suited to individual needs.

The experienced needleworker might choose an approach to plant dyeing that differs drastically from that of the beginning spinner and weaver. Someone living in the city has access to reference material not available in the country. Such personal circumstances affect each dyer's basic concerns. Flexibility is important. Dyers are wonderfully adaptive and inventive people. Indeed, what can tax one's ingenuity more than the practice of a once-dying skill in a time when the world economy is based on planned obsolescence? It is no coincidence that a concern for recycling and an interest in plant dyeing often go hand in hand.

Making dyes requires time and effort. The process is 'successful' if the colours are attractive to the dyer and the yarn quality has not been impaired. However, this is only the beginning. Using small swatches of yarn to illustrate a sample book is one thing, but to never venture beyond this is to miss entirely the joy of plant dyeing. Tapestry weavers and embroiderers use much less yarn per colour in any project than do weavers who make hangings. Such craftspersons will use small amounts of harmonious shades. The person working on a large scale, whether in weaving, macramé, crochet, sprang, knitting, or hooking will require several pounds of each of the selected colours for a project. The initial concept of a large project is likely to be constricted if the dyer is able to produce only small amounts of a given colour. Contrary to another myth, it is not more difficult to dye large amounts of fibre, if you know how to proceed (see p 61). The same amount of human and electrical energy is expended, so think big!

THE DYEING PROCESS

1 Select the dyestuff to be used. Gather and process according to type (flowers, whole plants, leaves, barks, roots, berries, lichens).
2 Select fibre to be dyed and prepare according to type (eg, yarn, fleece). Always soak out any fibre to be dyed for at least one hour – the longer, the better.
3 Determine which mordants (if any) are to be used (see pp 23–7), and which type of pot (see pp 10–12).
4 If desired, pre-mordant the fibre (pp 31, 35).
5 Place the dyestuff in the dyepot, add water to cover, and bring to a boil (212°F, 100°C). Remember, it is all right to boil the dyebath before the fibre has been added. Cook until the pigment has been extracted from the dyestuff (½ to 3 hours, depending upon the plant material used), or the bath appears to have sufficient colour. (Dip some in a glass measuring cup and hold it to the light.) NOTE: some very potent dyebaths appear deceptively pale at this point, but once a mordant is added they 'come alive.'
6 Strain the liquid (now called the dyebath or dye liquor) from the cooked out plant material, discarding the latter in a safe place.
7 Add additional water to the dyebath, if necessary, so the fibre will be able to 'swim freely' in the bath. (See Table 4, p 63.) Adding additional water will not decrease the strength of the bath but will ensure more even dyeing (see p 37). Have the pot just full enough so that there is no danger of the dyebath boiling over. Handspun and thick yarns tend to expand more than thin yarns and so require slightly more water in the bath.
8 If using the simultaneous mordanting technique (p 36), dissolve the mordant in boiling water and add it to the dyebath, stirring well.
9 Now add the wet fibre to the dyebath. Make certain that the fibre is approximately the same temperature as the bath. If the bath is hotter than the water in which the fibre has been soaking, add cold water to the bath to cool it. Or, conversely, the fibre may be placed in another, hotter rinse bath to raise its temperature to match that of the dyebath. However, controlling the temperature of the dyebath by adding additional hot or cold water is more satisfactory than attempting to increase or decrease the temperature of the wet fibre.
10 Start heating the dyebath to a slow simmer (190°F, 88–90°C) over the period of an hour (only thirty minutes or so for soft yellows; see p 65). Remember that now the fibre is in the pot, the dyebath must not boil. Maintain this temperature for one-half to three hours or more, depending upon the fibre being dyed, the dyestuff used, and the colour desired. (Long cooking produces brighter but darker colours.)

11 During dyeing, turn the fibre over in the bath and move it around to ensure that the dye penetrates easily. Add additional water to the bath if necessary. To do this, first remove the fibre, then add the water and replace the fibre in the pot. There should be sufficient dye liquor in the pot at all times so that the fibre 'swims' easily in it and is not jammed down against the bottom of the vessel.
12 To mordant at the end of the dyeing (see saddening and blooming, p 39) remove the fibre from the dyebath and add the dissolved mordant. Stir well, and return the fibre to the pot. Process an additional ten to twenty minutes, or until the desired colour has been obtained. Turn the yarn over to ensure thorough penetration of the saddening or blooming agent (iron or tin).
13 The dyeing process is finished when the desired colour has been reached. Remove the fibre from the dyebath or, if preferred, leave it in the bath overnight to further intensify the colour.
14 The dyebath may now be used a second time as an 'exhaust' bath, and even a third or fourth, if sufficient pigment remains. The *Journal of the Chicago Horticultural Society* 40 suggests combining the two exhaust baths for interesting effects such as a yellow and a red. Otherwise, discard the bath in a safe location. Dyers with septic tanks should dispose of their baths by pouring them in a ditch or even on a manure pile – anywhere – but away from children and pets. City dyers can pour them down the drain; the level of pollution is such that your dyebath will not add significantly to the problem. If your conscience bothers you, lug the dyebaths off to the municipal disposal plant.
15 Rinse the dyed fibre in successive baths of clear water, starting with a bath equal in temperature to the temperature of the dyebath itself, and finishing off with a cool rinse. (If the fibre is left to sit in the dyebath overnight, all the rinses are in cool water.) There should be three or four rinses, each slightly cooler than the last. Too much rinsing is better than too little. Thorough rinsing prevents 'crocking' (see p 66) and increases the fibre's resistance to fading from repeated washings. There will be some colour run-off during the rinsing, but this is normal. The addition of 1 Tbsp (15 ml) of vinegar will prevent excessive run-off, but may also slightly alter the colour of the dyed fibre.

SAMPLE ONION SKIN DYEBATH

Using the above procedures, here are step-by-step directions for making an onion skin dyebath sufficient to dye 8 oz (226.5 g) of 2-ply 'natural' white wool yarn (see p 42).

1 The dyestuff consists of 8 oz (226.5 g) of onion skins, an amount equal to the weight of the fibre being dyed. This produces a medium strength dyebath (see p 23). The skins are placed in the dyepot with water to cover (see water chart, p 63). Allow the skins to soak out for half an hour or more.
2 The fibre is white wool, in 3×4-oz (114-g) skeins. One skein has cotton string ties, another linen ties, and the third acrylic ties, to designate the mordants and dipping sequence (pp 34, 54). The skeins are soaked out by placing them in a sinkful of warm water, where they remain until the dyebath is ready.
3 As onion skins are a substantive dyestuff (see p 22), they do not require a mordant. However, the use of mordants greatly extends the colour possibilities. Those being used, therefore, are alum, cream of tartar, and tin, for blooming. The dyepot (which the skins are soaking out in) is enamel.
4 The fibre has not been pre-mordanted. It will be mordanted in the pot by the simultaneous method (p 36).
5 Put the covered dyepot containing the water and onion skins on the stove to be cooked out. Raise the heat until the dyebath reaches the boil (212°F, 100°C). Remove the pot lid occasionally and stir the skins down. The colour will be released into the dye liquid when the skins appear completely wilted and the dye liquor has a good colour, probably a rich rusty brown, depending upon the type of onions the skins were taken from (see p 23). It will take approximately half an hour for the skins to cook out, from the time the bath reaches a boil. If the colour appears satisfactory before then, fine. If it appears quite pale, then cook out the skins another 15 or 20 minutes.
6 Strain the liquid (which is now the dyebath) from the cooked out skins using a colander or straining cloth and a plastic pail. Discard the skins. Pour the dye liquor back into the dyepot.
7 Now add additional cold water to cool the dyebath. There should be enough water in the pot so that the fibre can 'swim' freely in the dyebath. For 8 oz of yarn, 3 to 4 gallons (13.5–18 l) is sufficient. Too much water is better than too little. (See Table 4, p 63.)
8 Dissolve the alum and cream of tartar in boiling water and add them to the dyebath, stirring well. For 8 oz of fibre, 2 Tbsp (30 ml) of alum and 2 tsp (10 ml) of cream of tartar are used (see mordant usage table, p 29).
9 Now add the 2 skeins of the wet fibre to the dyebath. If the fibre seems cooler than the bath, warm it by rinsing the yarn in warmer water, or, cool the bath by adding a little cold water to it.
10 Start heating the dyebath to a slow simmer (190°F, 88–90°C) over an hour. The pot can be covered or not, depending upon how full it is (a

very full pot is dangerous to cover as it may boil over). Make certain the fibre is completely covered by the dye liquor. If water evaporates noticeably during the processing, remove the yarn, add more water, and then put the yarn back in the pot. This will not dilute the strength of the bath (see p 63). Check the colour occasionally by lifting the yarn out and holding it above the pot with a stirrer. It is 'done' whenever it pleases you, the dyer. Keep in mind that yarn is a darker hue when wet than it is dry.

11 During the dyeing, move the fibre around to expose it all to the pigment and mordants.

12 When the desired colour has been obtained, remove both skeins from the dyebath. Put one aside in a plastic pail, without rinsing it first, and then proceed to rinse the other skein. Dissolve the tin (¼–½ tsp, 2.5–5 ml) in boiling water and add it to the dyebath. Return the unrinsed skein to the dyebath.

13 Process this skein for 10 or 15 minutes. Keep the pot uncovered so that it does not overheat, maintaining a temperature of slightly less than 200°F or 95°C, or turn off the heat: see p 40.

14 Rinse this skein and then discard the dyebath unless there is sufficient pigment for an exhaust dip. If the dye liquor still has colour, enter the third (as yet unprocessed) skein. So that you will have a third skein in case you want an exhaust bath, make a habit of soaking out more yarn than you expect to dye. This way there is always some yarn ready for an exhaust bath. Mordants may or may not be added to the exhaust dip, but if they are, they should be used in lesser amounts than those recommended on page 29. Having added the tin to bloom one skein, it might be interesting then to add a little iron.

15 Rinse the exhaust-dipped skein. You now have three different colours from a single bath: a rich, strong shade of medium yellow (depending upon the type of onions used) from the alum and cream of tartar; a bright orange or rust from the tin blooming; and, if you used the exhaust bath, a soft brown (or, if iron was added) a brown-olive shade.

DYEING LARGE AMOUNTS OF YARN

Because cramming too much fibre into a dyepot will result in uneven dyeing, dyers wishing to process several pounds of yarn in the same dyebath can use either of the following methods. The disadvantage of the copper boiler method is, of course, that the colours obtained using this utensil are always going to be in the same range (bronze, khaki, green, dull olive). However, the method is excellent for getting medium to dark browns, especially from onion skins and the lichen, *Lobaria pulmonaria*. (A good dark brown was obtained using the copper boiler method with two pounds of skeined natural white wool and one pound of very dark-skinned home-grown onions.)

Copper Boiler Method

Because of its enormous size (about 10 gallons, or 45 l) a copper wash boiler can easily hold two or three pounds of yarn for dyeing. Have all skeins weighing 4 oz (114 g) each, however, as this size will allow the dye to penetrate more fully than if the yarn were in heavier skeins. Cook out the dyestuff and strain it off. Pour the dye liquor back into the boiler, and add the dissolved mordants if using the simultaneous method. Simultaneous mordanting is the most satisfactory method to use when dyeing large amounts of fibre using either method 1 or 2. This gives the dyer a greater degree of control over the resulting colour and eliminates more variables which might influence the results. A stainless steel trivet can be placed on the bottom of the dyepot to prevent the yarn from streaking where it comes into contact with the hot bottom of the boiler. Have the fibre well soaked out in advance, preferably overnight. Add the yarn to the bath all at once so that some skeins do not absorb more colour upon contact than others. Turn the yarn over frequently during processing to ensure even dyeing. Keep the pot uncovered and the water level as high as possible to provide maximum room for the skeins. Process longer than normal, from one to two hours, and leave the skeins in the dyebath overnight for extra colour fixation. Rinse the following day. NOTE: Some copper boilers are tin-lined. Yarns dyed in such a vessel will have a characteristic tin-mordanted quality; that is, the colours will be bright and sharp-looking.

Three Pots at Once Method

For this method to work successfully and produce the same shade in each pot, dyeing vessels must be the same size and made of the same metal. (Dyers can use two, four, or even five pots – as many as the stove will hold at the same temperature.) If one pot is larger than another, it is difficult to estimate how much dyestuff and fibre it should contain unless the dyer has accurate scales to weigh the plant material and the fibre. The dyestuff must be collected from a single location. For example, picking goldenrod from one field one day and another field another day will not work satisfactorily unless the dyer knows that the soil in both fields is identical. Such esoteric information is more significant to the agriculturalist than the craftsperson or dyer, so pick the dyestuff for this method on a single outing. Cook out the dyestuff, using as many pots as required, and strain off the dye liquor. Use the same type of pot for each batch, and make sure each is perfectly clean and free from rust or stains. Wet the skeined fibre beforehand. Now divide the dye liquor, by volume, into the dyepots. Each pot should contain the same amount of dye liquor and the same weight of fibre. Have someone help you place the fibre in each pot at the same time. Slight inconsistencies in heat will not affect the results, but each pot should be processed the same length of time. If using a wood stove, move the pots around so that they receive equal heat. Process the baths using the simultaneous mordanting method. This enables the dyer to control the mordanting better and reduce the variables in processing. Cook for one to two hours, or until the desired shade has been reached. Leave the fibre to cool in the baths for extra fastness if desired, and rinse well the following day. The dyer should strive to eliminate as many variables as possible using this method. Use pots you are familiar with, a predictable dyestuff, and fibre you know and understand. Use of a new yarn may produce a colour that is unexpected and inappropriate for the project planned. Save experimentation for another time. NOTE: The pots used for this method must be absolutely clean. Rinse with chlorine bleach (see p 15). A minute amount of a mordant such as iron remaining in one pot could mean that the fibre processed in that vessel would dye quite a different colour.

Yarn Weight and Water Volume Table

Fibre to be mordanted or dyed requires sufficient liquid in the bath ('bath' refers to both the mordant and dyebath) to enable the yarn to 'swim freely.' This is essential for even mordanting and dyeing. Fibre processed in too little water may streak, appear blotchy, or even shrink. It is always safer to use more water than the recommended amount rather than less. If the mordant or dyebath is too full, pour a little out. If there is not enough liquid in the bath, remove the fibre and add more. Table 4 suggests minimum amounts of liquid required for processing various weights of fibre. Generally speaking, handspun and very thick yarns (5- or 6-ply) require slightly more water in the bath.

INCREASING AND DECREASING DYEBATH STRENGTH

Read definitions of a weak, medium, and strong bath on page 23.

In order to increase the strength of a dyebath (for a brighter, stronger, or darker colour), you simply increase the amount of dyestuff used to make the bath. Say the dyestuff is fresh lettuce leaves, and you have prepared a weak bath, using one part lettuce to two parts fibre (by weight). The wet fibre has been processed half an hour, and the colour

Table 4: Minimum amounts of liquids required for dyebaths

Weight of fibre to be processed	Volume of liquid
1–2 oz; 28–57 g	1 gallon; 4.5 l
4 oz; 113 g	2 gallons; 9 l
8 oz; 226.5 g	4 gallons; 18 l
1 lb; 453 g	6 gallons; 27 l
2–3 lbs; approximately 1 kilo	10 gallons; 45.5 l

appears too pale to your liking. Make a notation in your dyeing sample book to use a medium strength bath next time. You could, at this point, remove the fibre and collect more lettuce. Process it in another pot, strain off, and add this additional dye liquor. Also, the use of certain mordants may produce a colour you find more attractive. In this case, perhaps blue vitriol or iron would effect a change.

Some dyestuffs contain more pigment than others. Say you have made up a bath using fresh maple bark, collected from a felled tree. The bath was of medium strength (2 parts bark to 2 parts fibre, by weight), and after one hour of processing, you find the colour is much darker than you would have wished. Aside from removing the fibre then and there, there is not much you can do this time. However, a notation to use a weak bath next time should go in your dyeing sample book. And you will undoubtedly remember that this particular dyestuff is quite strong. A medium strength bath of apple bark would probably have given a good light brown or tan.

The addition of water to a dyebath will not affect its strength (p 37). Once the dyestuff has been cooked out, and the liquor strained off, the amount of pigment released into the water or dyebath is constant. It can be changed in colour (using a mordant), but only diluted in potency by using more fibre than the amount planned. Before the fibre is placed in the dyebath, hold a glass cupful of the bath to the light. If it appears much darker or stronger than anticipated, soak out additional fibre and dye more yarn.

A dyebath is more often too pale than too strong. Dip a glass cupful and hold it to the light. If the colour is much paler than you expected, simply enter less fibre in the basin. If you planned on dyeing 8 oz (226.5 g) of yarn, and prepared a medium bath, then use only 4 oz (114 g) instead. Some pale-looking baths are quite deceptive and turn a dramatic colour once mordants are added. This is the case with fresh Juneberry leaves (see p 160). The bath was a soft yellow-tan, but the addition of tin pre-mordanted yarn resulted in a brilliant orange-rust shade. Lichen baths react similarly, often appearing much paler than they actually are.

PROCESSING TIME

'Processing time' refers to the duration of the whole dyeing procedure, from the cooking out of the dyestuff to the final rinsing of the fibre. The processing time varies, depending upon the fibre being dyed, the type of dyestuff used, the method of mordanting, and the colour the dyer hopes to obtain. Generally speaking, the longer a bath is processed, the brighter, richer, and/or darker the resulting colour will be. For example: 8 oz (226.5 g) of fresh blackberry shoots (see p 105) soaked in water to cover overnight, and then cooked out the following day for two hours, will yield a more intense shade than the same weight of shoots not soaked overnight and cooked for only half an hour. Once the dye liquor has been strained off and the dyestuff discarded, the timing of the actual dyeing (and mordanting, if using the simultaneous method) begins. Dyeing/mordanting for one hour, then, means having the fibre in the dye/mordant bath for one hour total time. This includes the time it takes for the pot to reach a simmer, but does not include the time it takes to rinse out the fibre after dyeing. Most baths are processed for two or three hours, starting with cooking out of the dyestuff and ending with removal of the dyepot from the heat. The actual dyeing time varies, but most baths require forty-five minutes to an hour. Exceptions are tannin baths (below), and yellow baths from flowers (p 65). Lichens are often cooked out for three hours prior to dyeing (Bolton, *Lichens for Vegetable Dyeing* 53). Soft shades require weak or medium baths and short processing, while strong and dark colours require a strong bath and long processing. This is most evident with a dyestuff such as onion skins. If two baths of equal strength (say, medium) are processed for a different length of time, the colour of the one receiving the longer processing will be decidedly different from the other bath: the short bath might be yellow (alum mordant), while the long bath could be an old gold (same mordant).

TANNIN BATHS

Baths made from acorns, nuts, barks, and roots require a maximum processing time to release their pigment (see dyeing with bark and roots, p 73). Such dyestuffs yield their colour more readily if they are soaked out in water to cover 24 to 48 hours prior to the cooking out. However, as previously noted, dyestuffs containing tannin may produce a dull, dark brown rather than a warm, rich brown shade (see p 40 for a discussion of tannic acid). If the dyer wishes the latter colour, then bark and roots should be processed at a temperature well below the boil (maximum 190°F, 88°C) to avoid the extraction of an excessive amount of tannin (Jo Lohmolder, 'Dyeing with Bark,' *Shuttle, Spindle & Dyepot*, Issue 18, V, 2 [Hartford, CT: Handweavers Guild of America 1974] 76).

YELLOW BATHS FROM FLOWERS

An exception to the general rule for long processing to intensify a colour is the yellow dyebath made from fresh flowers (daffodil, calendula, zinnia, marigold, and so on). If the dyer wishes a bright but clear yellow (and not a gold or bronze), then the flower bath must be processed for less than forty-five minutes with the temperature kept at or below 190°F or 88°C. This will produce a clear, soft yellow. To brighten the colour, the bath is bloomed with tin (see blooming, p 39).

SAVING AND STORING DYEBATHS

Dyers who demonstrate or teach their craft during the winter have devised methods for saving dyebaths for future use. The baths are freshly prepared when the plants are in season. Freezing: the dyestuff is covered with water and cooked out. The dye liquor is then strained off and poured into suitable containers for freezing. Plastic containers are best. Never freeze dyebaths in metal containers. Glass jars will do if the liquor is cool beforehand. Leave an inch of space at the top of the container to allow for expansion during freezing. Evaporation: the dyestuff is cooked out in water to cover, and the dye liquor then strained off into a clean container. The liquor is left in a warm place until all the liquid evaporates, leaving the concentrated dyestuff in powder form. This method is also used to reduce fermenting orchil lichen solutions to a powder (Bolton, *Lichens for Vegetable Dyeing* 46). Refrigeration: Robertson recommends storing unmordanted baths in the refrigerator, but I have found these will mould after three to four weeks (Robertson, *Dyes from Plants* 20). Chemical preservation: Lesch gives directions for storing baths to which sodium benzoate has been added as a preservative (Lesch, *Vegetable Dyeing* 16). The recommended amount is 1 tsp (5 ml) to 1 gallon (4.5 l) of liquor. The mixture is then stored at room temperature in airtight, non-metal containers. Another method commonly used is to store the dye-stuff itself by drying or freezing it (see dyeing with flowers, p 69). Onion and flower baths are often stored at room temperature until they mould, to produce rich colour effects. The bath may just be steeped, or actually cooked out first (Robert and Christine Thresh, *An Introduction to Natural Dyeing* 34). In any event, make it a practice not to store, by any method, a dyebath to which mordants have been added. Toxicity may result, and increase upon extended storage, causing harm to the dyer.

DYEING NON-WHITE YARNS

Brown, grey, and beige yarns may be dyed as readily as white. Actually any colour of fibre can be dyed, with the resulting shade depending upon the initial colour of the yarn itself. Obviously, grey yarn will not dye a pale or soft shade of clear yellow or mauve, but it will pick up interesting highlights which give it an attractive 'heather' look. At a dyeing workshop at the Nova Scotia College of Art & Design, Halifax, in 1976, medium grey wool yarn was dyed in an *Umbilicaria* bath, mordanted with chrome. The result was a lovely purple. Chrome-mordanted medium grey yarn in an onion bath dyed a greyish-green, with the addition of iron. Brown yarns in an onion bath take on a rich, golden cast, and grey yarns in green baths become a dark greyish-green. There are numerous possibilities which experimentation will turn into favourite standard recipes in a dyer's repertoire.

STREAKING AND TIE-DYEING

Yarn that streaks in the mordant/dyebath may have been tied too tightly (see p 54). However, if the dyer wishes to produce a tie-dyed yarn, and to incorporate that effect into a project, the fibre to be dyed is tied too tightly on purpose. However, do not allow yourself to slip into the habit of saying you meant a fibre to streak when, in fact, you did not. Legitimate tie-dyeing is not an excuse for inadequate dyeing skills. To tie-dye a fibre, use ties of a strong cotton or linen twine or cord. Tie the skein (if the fibre is yarn) in as many places as you wish, wrapping the tie as tightly as possible around the skein. Do not divide the skein or make figure eight ties. Plastic wrap may be placed either under or over the ties, to make the wrapping even tighter. Some dyers use metal twist ties for tie-dyeing, or elastics. Wherever the skein has been bound up, the mordant/dyebath will not penetrate, but the exposed portion of the skein will dye as usual. However, because fibre absorbs colour so readily, a little colour will probably penetrate the extremities of the bound sections, producing a delightful sun-ray effect.

FASTNESS AND CROCKING

It is important that fibres used for garments be fast to light and washing. It is equally important that all plant-dyed fibres respond to exposure to sunlight and water in a known way before the dyer uses them in a project. To weave a rug of plant-dyed wools and then have it fade appreciably would be disappointing as well as a waste of time. Similarly, colours that 'crock,' or rub off onto an adjacent fibre, will spoil an entire piece of work. Fading and crocking can be prevented if the dyer processes the fibre methodically, making certain that the rinsing is thorough. Rinse, rinse, and rinse some more. Then test the dyed fibre as follows: Take two pieces of cardboard and cut a 'window' in one. Insert a small hank of dyed yarn between the cardboard to make a 'sandwich.' Allow some of the fibre to be exposed to the light through the cut window, and then tape up the edges of the sandwich with masking tape. Expose the sandwich to the sun over a period of ten days to two weeks (it may take longer in the winter). Then remove the cardboard and record the results in your sample book. If the fibre has faded noticeably, and is therefore not very fast to light, do not use if for rugs, hangings, cushions, curtains, or garments. Take a small piece of yarn from the same skein you have tested for fading and test it for washability. Wash and dry it and then compare the colour with that of the rest of the skein. Record the results. To test for crocking, rub a dry piece of the yarn over a white cloth or rag (cotton or linen). If the pigment rubs off onto the white cloth, do not use that fibre for a project where this would affect the article's washability. It might do for a hanging, but not for upholstering the family's favourite television chair. Crocking may sometimes be controlled by adding vinegar to the final rinse after dyeing. However, the acidity of the vinegar might also alter some colours.

It can be said that all plant-dyed fibres fade to some degree. Indeed, most commercially dyed fibres fade as well, but sceptics fail to point this out. One summer I satisfied myself that plant-dyed fibres fade no more than other fibres. I had tied red polyester and cotton cloth ties to my fruit trees, which are adjacent to the hay fields, to identify the trees to the mower and prevent them being accidentally cut down. When the haying was over, two weeks later, I decided to leave them on the trees to save the cost of buying more broadcloth the next summer. All the ties had faded to near-white. A few were pale pink. This fading occurred during a two-week period when the sun was, admittedly, strong. If polyester cotton can fade, so will your plant-dyed fibres, but that fading is in fact a mellowing. A yellow-green may change to a light avocado; a strong rust to a burnt orange; gold may become ochre. As is true with every other aspect of plant dyeing, the beauty of the craft lies in just such subtleties. Learn to flow with nature, rather than fighting it, for a truly comprehensive understanding of plant dyeing. It is the mark of an accomplished craftsperson to produce not only a fine object but also contribute skills to an already growing body of knowledge that benefits all who work with their hands.

SAFETY PRECAUTIONS

1 NEVER USE DYEING EQUIPMENT FOR FOOD STORAGE OR PREPARATION. Keep all utensils out of the reach of children. Keep chemicals labelled POISONOUS under lock and key or where no one else will confuse them with laundry supplies. Assume responsibility for all the equipment and plants you use in dyeing.

2 Do not mordant or dye with alum, blue vitriol, chrome, iron, or tin in a closed room where there is no air circulation. Do not mordant or dye with these chemicals when infants are in the same room.
3 Dispose of all mordanting and dyeing refuse where pets and children cannot get into it.
4 Make a habit of using rubber gloves all the time. Do not serve or eat food while mordanting or dyeing.
5 Wash all countertops well after mordanting and dyeing, and wipe up floor spills with ammonia and water or bleach and water.
6 Use dyestuffs that require long processing on a day when windows and doors can be kept open. Use a kitchen fan or exhaust if you have one.
7 Always tell someone else in the household what you are working with. Explain the mordants being used, and the dyeplants, especially if working with mushrooms. Identify a mushroom sample, using the correct Latin name for the species and genus (eg, *Agaricus campestris*). Leave this sample to one side while you work.
8 Always cover a mordant or dyebath which is left sitting for any period of time. If a bath is to be left in one place for several days, tie the lid on with cord. In the summer, leave baths outdoors in a safe place if they are to sit for some time; in the winter, leave the bath in a room that has some ventilation or air circulation. Again, tie on the lid, especially in a family where there are small children and pets.

Most of these safety precautions are self-evident, but the beginning dyer is advised to be safety conscious from the outset. Establish good dyeing habits and stick to them. Careless dyers may harm themselves and others; it pays to realize you are working with POTENTIALLY DANGEROUS substances. But then, so does the cook. Just as we learn to handle sharp knives and refrigerate bacteria-prone foods, so we can learn the elements of safe dyeing to maximize the pleasure it gives and minimize the risks.

7

The dyestuffs

Just as fibre must be appropriately prepared to receive a mordant or dye, so must the actual dyestuff be processed to release its pigment. Dyestuffs are processed according to their type: fresh flowers, fresh leaves, whole plants, lichens, mushrooms, barks, roots, nuts, and berries. The method of extracting the pigment from each group of dyestuffs varies slightly, but essentially involves soaking the dyestuff in water to cover and then cooking this mixture until the pigment is released into the water. This water then becomes the dye liquor, or dyebath.

Beginning dyers are often confused when they discover, by reading or experimentation, that the colour of a dyestuff does not necessarily indicate what colour the dyebath made from the plant material will be. True, onion skins give a range of colours from yellow through to brown and khaki. It is also true that most yellow flowers give yellow to yellow-green baths. However, pink field clover gives a bright chartreuse, and purple lupins a lovely lime green. Not all blue flowers make a blue bath; a few will, including delphinium, but the colour is hardly a true blue and rarely fast. The spice turmeric gives a brilliant yellow but it is fugitive and will wash or fade out. But the fresh green leaves from wild apple trees will make a fine rust (fresh leaves, a medium bath, with blue vitriol and chrome as mordants) and apple tree bark gives a soft yellow of unequalled beauty. Remember to keep a sample book and record these fascinating disparities for yourself (see p 84).

DYEING WITH FLOWERS

Fresh flowers are processed according to their species and colour. Common baths are yellow marigolds, orange marigolds, rust marigolds, yellow calendula, blue delphinium, pink clover, goldenrod heads, Queen Anne's lace, white lupins, pink lupins, and purple lupins. Most flowers give the strongest and brightest shades when they are picked just prior to being in full bloom. The flower tops or heads are shredded or torn apart with the hands and placed in the dyepot. Cover the flowers with water, pushing them down until all are well soaked. The bath is then cooked out for the desired length of time (see processing time, p 64, and yellow baths from flowers, p 65). The flowers are strained off and discarded. The wet fibre is entered. If it has not been mordanted (see pre-mordanting, p 35), then the mordants to be used are dissolved and added at the desired time. This depends upon whether the bath is to be mordanted by the regular simultaneous method (see p 36) or bloomed or saddened (see p 39). If the bath is to be mordanted by the regular method, the mordant is added before the wet fibre; if the bath is bloomed with tin or saddened with iron, the chemicals are added at the end of processing.

If stems and leaves are included with flowers in the dyebath, resulting colours will be greener and darker. For clear, soft yellows, keep the temperature below a simmer (200°F, 95°C). Flowers collected in the rain, or after a heavy dew, seem to cook out more readily. Some varieties, such as marigolds and calendula, give richer, deeper colours when used for a dyebath after they have been touched by frost. Flowers that have been frozen, or dried and saved for later use, may give a somewhat different colour in the dyebath. Cape Breton dyer Eveline MacLeod says she found using frozen marigolds resulted in a darker shade. One reference advises against freezing blooms but uses the air drying method to preserve them (Lesch, *Vegetable Dyeing* 15). Blooms dried this way tend to produce paler shades.

Many dyers new to gardening may not have enough of a species of a plant to use that flower for a dyebath. Many varieties of flowers can be combined in a single dyebath, and such baths often give interesting results. The following are some often-used combinations: day lilies and yellow iris; daffodils and tulips; pansies and petunias; asters and zinnias; peonies and roses; chrysanthemums and coreopsis; coreopsis and sunflowers. Flowers may be combined by their colours: whites and pinks, yellows and golds, oranges and rusts, reds and purples, blues and purples, bronzes and golds, and so on. Although flower colours do not indicate resulting dye shades, similarly pigmented blooms do seem to have an affinity for one another in the dyepot. Baths using a mixture of colours according to species – for example, mixed pansies, petunias, or begonias – tend to produce a grey colour.

Faded blooms and petals are also used for dyebaths. Again, the species may be mixed or not, according to what is available. Likewise, colours can be mixed. The blooms are picked just as they fade and are no longer attractive for ornamentation. Separate the petals as much as possible, and place them in a plastic container with water to cover. In order to allow air to circulate, do not cover it. As soon as enough petals have been collected, make a dyebath in the usual manner. The colours produced are extremely bright and attractive, although the bath is foul-smelling. A dash of baking soda can be added to the soaking petals to eliminate or, more accurately, modify the ensuing smell. As previously mentioned (p 65), flower baths are often left to soak for an indefinite period of time, both before and after they are cooked out. This extended soaking produces excellent results, but the odour is substantially more noticeable than if the flowers are freshly processed.

DYEING WITH LEAVES

For the purpose of this book, all types of plant foliage are considered 'leaves.' Carrot tops, spinach, purslane – all are leaves and are processed as such. However, if the bloom, leaves, and stems are used as the dyestuff, as is the case with plants such as bachelor's buttons and sorrel, then the plant is processed as for 'whole plants.'

To obtain soft yellow-greens from fresh leaves, they should be picked just as they reach maturity and used forthwith. Mordants such as alum, blue vitriol, and iron tend to enhance the 'green' range thus obtained. As with flowers, picking leaves on a wet day, or after a heavy dew, makes them easier to cook out. Like flowers, leaves are shredded first or else torn apart with the hands. Because leaves are slower to cook out than most flowers, they benefit from soaking out overnight in water to cover in the dyepot. Yellow-green baths from fresh leaves may turn to beige and tan if processed at too high a temperature. Keep the bath below a simmer (200°F, 95°C) and make certain no twigs, nuts, or bark are clinging to the leaves after they are picked, as the tannin in such substances would turn the bath away from yellow-green to tan and brown.

Coloured fall leaves may also be used for a dyebath. The most interesting results are obtained using leaves such as blackberry and raspberry. Leaves from these wild fruits turn scarlet to purple in the fall, and both produce fine browns and greys. When collecting fall leaves for a dyebath, make certain they have fallen on clean ground. Leaves contaminated by road dust, smoke, and dirt will produce quite nondescript murky shades.

DYEING WITH WHOLE PLANTS

Using all of a plant in the dyepot is quick, convenient, and less wasteful than merely picking off the blooms, especially if they are small (eg, buttercup, violet). A variety of common cultivated and wild plants lend themselves to this type of processing, including goldenrod, dock, wild mustard, and lamb's quarters. Whereas processing only the blooms of goldenrod gives a yellow-gold-bronze bath, using the whole plant will result in an avocado or khaki shade, depending upon the mordants used. The plant for the dyestuff is first torn or shredded. The root may or may not be left on. If it is left on, make certain all dirt clinging to it is brushed off, otherwise the resulting bath will be murky and dull. Although roots are, in themselves, useful dyestuffs, the roots of most plants are too small to collect as a single dye material. However, roots contain much pigment and often contribute an interesting colour influence when used along with the rest of a plant in the dyepot. Most whole plant dyeing results in shades of medium yellow-green, medium grey green, bronze, khaki, and avocado. If a plant to be processed whole is tough, like goldenrod, which has woody stems, it is beneficial to soak out the dyestuff in water to cover for at least twelve hours. A longer soak will be even better.

DYEING WITH LICHENS

Complete information on dyeing with lichens begins on page 164. Essentially, there are two techniques: the boiling water method and fermentation with ammonia. Most lichens are processed in boiling water, producing yellows, golds, rusts, and browns. They are cooked out after being thoroughly torn up and allowed to soak in water to cover for twenty-four hours. Species of the genus *Umbilicaria* and one of the *Parmelias* are processed with ammonia and yield outstanding and unequalled reds, pinks, magentas, and purples.

Dyeing with lichens is a unique experience. Once a dyer pulls a brilliant magenta-coloured yarn from the dyepot, there is no turning back. Lichens impart a characteristic smell to yarn dyed with them. While this is 'strong' to some people, notably non-dyers, it is attractive to the dyer, who knows and loves the fungus-algae relationship we know as lichens.

Disappointing results from lichens may discourage beginning dyers. Most lichens must be used in a strong bath (see p 76) to yield their maximum colour. They benefit from a long soak (even several days) but must be processed below a simmer (200°F, 95°C) to ensure colour clarity. Supposedly, lichens collected in late summer give stronger colours (Bolton, *Lichens for Vegetable Dyeing* 19), but I have tested *Umbilicaria* collected in all seasons and found no disparity in the results. *Umbilicaria* species collected were the same: *Mammulata* and *Deusta* combined; both baths were fermented the same length of time and mordants used in the dyeing were identical.

DYEING WITH MUSHROOMS

Collection of mushrooms for plant dyeing is dangerous unless the dyer has a basic knowledge of mushroom identification. This can be learned using one of the many excellent books available, but make certain the book contains those species which grow in the region where you live. A book of familiar northwestern mushrooms is no use! Most naturalist societies and museums offer mushroom identification courses, and dyers are urged to attend these as on-the-spot learning with the help of a botanist is extremely beneficial.

So, assuming you have this basic information, collect any variety which occurs in sufficient abundance. One fall the hundreds of chanterelles (*Cantharellus cibarius*) which grow in our woods were infected with some type of disease. Rather than let them go to waste, I used them for a very successful dyebath. Mushrooms that have spoiled (inkies, *Coprinus micaceus*; shaggy manes, *Coprinus comatus*) and cannot be eaten for that reason make interesting baths that produce greys with overtones of pink, purple, and green.

Mushrooms should be chopped up, if woody, or merely torn apart with gloved hands. Soak them in water to cover overnight, and cook out the following day. Strain off the resulting mush through something fine, such as cheesecloth. Mushroom baths seem to benefit from the use of chrome, iron, and tin. Alum and blue vitriol appear to have too little influence on the unusual colours from mushroom baths to be worthwhile, but you can always experiment and perhaps come up with something exciting.

Miriam Rice's *Let's Try Mushrooms for Color* (p 55) is excellent and informative, but few of the species she uses are indigenous to the northeast.

DYEING WITH BARK AND ROOTS

Bark and roots are tough and fibrous. A long cooking period is usually required to release their pigment. As noted on pages 40 and 64, dyestuffs containing tannin are apt to produce dark, dull colours unless processed at a temperature below a simmer (200°F, 95°C).

Early spring bark is the easiest to collect. It peels off twigs quite readily, although some dyers prefer to use the twigs unpeeled in the bark dyebath. Using the twigs of a tree rather than chunks of bark from the main trunk causes far less damage to the tree and results in the same colour. Inner bark has a greater concentration of pigment and is often specified in a dye recipe (Arnold and Connie Krochmal, *The Complete Illustrated*

Book of Dyes from Natural Sources 98). However, for the purposes of this book, any subsequent references to inner bark are based on the assumption that dyers using this will NEVER collect it from a living tree. Bark may be obtained from deadfalls, cut-off diseased limbs, and firewood. Logs cut for pulp are also a good source of bark. There is no excuse to endanger a healthy specimen. Some references suggest collecting bark at specific times of the year. Thurston, for example, advises collecting resinous barks (fir, pine, spruce) in the spring and other types in the fall (p 17). Furry and Viemont (p 8) recommend collecting non-resinous types (fruit trees, shade trees) in the fall and winter. I have collected and used all types at different times of the year and found no substantial difference in the dyebaths, although spring alder twigs produced a stronger and more interesting brown than those collected in the winter.

Roots are much the same as barks, except that they grow on that part of the tree which is underground. As such, roots are exposed to no sun, more moisture, and slightly different chemicals present in the soil immediately surrounding them. For this reason they may have more pigment than bark from the same species of tree, or produce a slightly different colour. It goes without saying that only roots from deadfalls or diseased trees are used. Clean off all dirt from roots before using them for a dye.

Barks such as apple and pear are easily removed with the fingers or a blunt knife. Birch-bark peels off and can be found in heaps at the bottom of living trees in the woods. Remove moss and lichens from bark if you wish to cancel the possible yellowing influence these plants might contribute in the dyepot. I always leave *Lobaria pulmonaria* on maple, ash, and oak bark as it produces a fine brown by itself, and can only serve to enhance the bark colour.

Chop tough barks and roots with a kitchen chopper, a small camper's axe, or, if in small pieces, the kitchen blender. It should be reduced to pieces no larger than 1 square inch (6.5 cm^2). Then cover with water and soak out for as long as possible. Three to four days is ideal.

To encourage the brown tannin effect, process the bath at a simmer. To avoid this, and aim for yellow-tans and warm brown, process below a simmer. For extra fastness, leave the dyed fibre to cool in the bath for twenty-four hours or longer. To ensure that the brown obtained does not darken with time, follow the bark or root bath with a dip in blue vitriol as suggested in the reference on page 40.

Oak, sumac, and alder galls may be used as dyestuffs, or as additives to other dyebaths to make black (Adrosko, *Natural Dyes and Home Dyeing* 50). Galls were traditionally used in tanning and the making of ink (Robertson, *Dye Plants and Dyeing* 8). They are round, black, crusty-looking growths caused by insects. The gall may be removed by cutting it off at its base or, if the entire tree limb seems diseased, remove the whole thing and scrape off the galls into a bucket. Often small trees such as alder and chokeberry are entirely infected, providing the dyer with an abundance of this material.

DYEING WITH NUTS

Acorns and other tree nuts make excellent dyes, and the green husks of the black walnut produce what is perhaps the best brown available to dyers. All nuts must be free from dirt and dust and then crushed with a hammer or mallet before soaking out. They benefit, as do bark and roots, from a long soak prior to being cooked out. Generally speaking, nuts are used in sufficient quantity to make a strong dyebath: that is, twice the weight of nuts to the weight of the fibre to be dyed (see pp 23 and 76). More details for dyeing with black walnut and butternut hulls are given under the listing of individual dyestuffs. Ida Grae (*Nature's Colors: Dyes from Plants* 188) gives interesting recipes for using walnut leaves to obtain brown and black, while Seonaid Robertson (*Dyes from Plants* 97) suggests top-dyeing walnut hull browns to get black.

DYEING WITH BERRIES AND FRUIT

Although many dyers such as Mary Frances Davidson (p 6) and Seonaid Robertson (p 34) have successfully documented their experience in dyeing with berries, others such as Margaret S. Furry and Bess M. Viemont do not consider berry dyebaths sufficiently fast to bother with them. Ida Grae ingeniously combines berries with flour and sugar in dyebaths, to more or less duplicate berry pie ingredients. Everyone knows that berry pie juice simply does not fade or wash out! My personal feeling about using berries for a dyebath is that fresh fruit is far too valuable a source of nourishment to squander it. But dyers who have a good supply of wild berries are encouraged to try them. Dyers who prefer to eat and cook with their berries, as I do, can still use the residue from jelly-making for a dye. Berries are best if picked when they are dead ripe. Keep the temperature below a simmer to preserve the pink tones and lessen the tan influence. Wetting the cloth through which you strain off the berry dyebath liquor saves more of the pigment for dyeing. Liquor poured through a dry straining cloth is absorbed much as ink when poured on a blotter.

Amount of Dyestuff to Use

As previously noted (pp 23, 76), the strength of a dyebath depends upon the proportion (by weight, or volume) of dyestuff to fibre in the dyebath. Baths are generally referred to as weak, medium, or strong. Each dyestuff

Table 5: Effects of weak, medium, and strong dyebaths

dyestuff	mordant	weak bath	medium bath	strong bath
fresh flowers, marigolds	alum	yellow	yellow-gold	old gold
fresh leaves, apple	chrome	warm tan	warm brown	rusty brown
whole plant, goldenrod	iron	yellow-green	olive to khaki	avocado green
boiling water lichen *Lobaria pulmonaria*	none	beige	tan	brown
fermented lichen *Umbilicaria muhlenbergi*		orchid	magenta	magenta-purple
mushrooms *Cantharellus cibarius*	blue vitriol	beige	tan	yellow-green
fresh apple bark	none	beige	yellow-tan	warm brown
black walnut hulls	chrome	brown	dark brown	brown-black
fresh chokecherries	tin	pinkish tan	mauve tan	taupe
tops from rhubarb (the leaves)	tin	yellow-orange	bright orange	rust
fresh rhododendron leaves	iron	warm grey	grey-green	olive-brown

Weak bath: one part DYESTUFF to two parts fibre (by weight or volume)
Medium bath: equal amounts of DYESTUFF and fibre
Strong bath: two parts DYESTUFF to one part fibre

will respond differently depending on how much of it is used. Various mordants will further alter, modify, or enhance the resulting colours. For example, a weak onion skin bath with an alum mordant will produce a different shade from a medium onion bath with the same mordant.

Table 5 shows the effect of a weak, medium, or strong bath using each type of dyestuff previously described (flowers, leaves, whole plants, lichens, mushrooms, bark and roots, and berries), with a variety of mordants. The table is simply a guide: depending upon where the dyer lives, and other variants, another person's results might well be different from those given here. Generally speaking, the most satisfactory dyebaths are those made using the largest amount of a dyestuff that is practical. Dyebaths are most often disappointing not because they produce too bright or too dark a colour, but because the colour is too pale.

8

Workshop and classroom techniques

WORKSHOPS

Plant dyeing workshops provide the beginning dyer with an excellent opportunity to learn techniques. However, unless it has been stated otherwise by the sponsoring group or organization, advanced dyers can expect to benefit only minimally from a day-long workshop geared to teaching the basics of dyeing. Workshops tend to vary in duration from several hours to a week or more in length. The more intensive the workshop, the more one can expect to learn. Longer workshops focus more attention on such matters as mordanting and the use of imported dyestuffs like indigo, procedures which may be too lengthy to fit into the schedule of a day-long event. Attendance varies, too. It may be as low as four or five persons at a workshop for experienced dyers, or as high as thirty at one for beginning dyers. Dyers seriously interested in upgrading their technical skills will find small, long-term sessions preferable to day-long events. However, one should take advantage of all that are offered if only to familiarize oneself with the approach of a different instructor.

BEGINNING DYERS' WORKSHOPS

Workshops for people new to plant dyeing are usually organized in such a way that the participant requires no previous experience in order to benefit from attending. The sponsoring group informs you if you need to bring your own fibre; usually this is provided and may be included in the workshop fee. Dyestuffs to be used are normally common ones, such as onion skins and various types of flowers. These may already have been collected by the instructor, or may be gathered during the workshop. Those workshops where the participants collect their own dyestuffs are invaluable in that they aid the dyer to learn plant identification. A lichen workshop for beginning dyers is not all that rewarding if it is not accompanied by an appropriate field trip. Participants are usually expected to take their own notes, asking questions about the techniques as the instructor demonstrates. If the workshop is fairly small, participants will probably do the mordanting and dyeing themselves.

ADVANCED WORKSHOPS

Advanced workshops seem to succeed best when the dyers in attendance are free to offer and exchange specific technical information. At such a session, the instructor may be more of a chairperson than a teacher. Topics are most often determined in advance, with the overall approach geared to the expressed interests of members of the group. Dyers with various levels of expertise can, in this way, provide each other with detailed information that might not even be discussed at a more general workshop. A small group of people wishing to study *Umbilicaria*, for instance, might meet for collecting at one time, and then reconvene later when their lichens have been fermented. The workshop would then resume with the actual dyeing of a variety of *Umbilicaria* baths, separated as to species and intensity of the dyebath. The comparative samples resulting from such a workshop would be very useful to serious dyers who, on their own, have not had the opportunity or motivation to test more than one type.

OUTDOOR WORKSHOPS

There is no doubt that an outdoor workshop is exhilarating and invigorating, but the weather can, and does, interfere. Rain, fog, and high wind, or hot sun can easily dampen enthusiasm and make everyone anxious only to go home. Driftwood fires are difficult to maintain at a consistent temperature. Pots processed on such a fire get red-hot, including their handles. This makes removing them from the heat quite difficult unless they are protected with a wrapping of aluminum foil. An outdoor cement or brick barbeque is fine to use, although most accommodate only one pot at a time. Outdoor dyeing can be good fun, but for serious work it is better to be indoors where you have the convenience of running water, several sources of variable heat, counter space, and comfort.

INDOOR WORKSHOPS

All indoor workshops should be held in an appropriate space where the following are available: fresh water, fresh air, several sources of heat, and sufficient room to allow each participant to work comfortably. There should be extra pails and buckets and enough rubber gloves to protect everyone who is actually mordanting or dyeing. All mordants should be in screw-topped jars bearing appropriate labels. Participants should be advised of their toxicity. Smoking, drinking, and eating in the working area should not be allowed.

WORKSHOP SAMPLES

Ideally, each participant should take home a sample of each dyebath processed. This may be a small hank of yarn or an entire skein. Because the fibres dyed are still quite wet at the end of a day-long session, the instructor has to untie a wet skein in order to give everyone a piece. A more efficient method is for the instructor or sponsoring agent to mail samples from each bath in stamped, self-addressed envelopes later. At advanced workshops, dyers are usually responsible for collecting their own samples from each skein. If they have dyed their own skeins, they bring plastic bags in which to take them home. The same mailing system, however, can enable the dyers to have samples of other people's skeins as well.

PLANT DYEING AS A CLASSROOM PROJECT

Plant dyeing is a fascinating classroom project suitable for children of any age, provided the teacher exercises care regarding the source of heat and material used, such as mordants. Students up to the age of seven or eight enjoy watching fibres being dyed, while those from nine to twelve derive the greatest satisfaction from actually participating in the process. Older students benefit from relating the dyeing experience to another subject, such as science. Projects suited to this age category might include the documentation of traditional dyestuffs of native cultures or the investigation of specific botanical forms, such as lichens.

Materials and Equipment

No extra space is required for dyeing in the classroom, but the amount of equipment used can vary depending upon the ages of the children. The younger the class, the less equipment you should have for the dyeing. This helps cut down on the risk of accidents. (Read safety precautions, p 67.)

- one or more enamel basin, pot, or pan (the larger the better)
- source of heat that is as sturdy as possible (hot plate, propane camp stove)

– fresh water (in large plastic jugs or from a nearby tap)
– rinsing basins or plastic pails (two for each dyepot used)
– rubber gloves (enough to provide each dyer with gloves, although gloves are not necessary if salt, vinegar, soda, or urine are used as mordants, ALWAYS use them if older students are working with ANY chemicals)
– sponges, paper towels, old rags, newspapers for cleanup
– strainer (plastic; a metal strainer or colander would react with the mordant) – mordant (vinegar, urine, salt, or baking soda)
– dyestuff, preferably substantive (see p 22), such as onion skins, marigolds. Other suitable plants are dandelion, pink clover, lupin, fresh leaves from apple, pear, poplar, or alder trees. In winter, get scrap foliage from a florist's shop, or use discarded tea bags and coffee grounds.

When choosing a dyestuff, allow the students to make as many decisions as possible. Explain substantive and adjective dyestuffs, and what a mordant does to the dye. Although onion skins and marigolds probably give the most visually exciting results, most everything will produce some colour. Young students respond more readily to bright shades, but older ones will appreciate the subtlety of less dramatic results, especially if they are students with some art background.

Procedure

NOTE: Read pot as mordant, page 38, and sample onion bath, page 58. If you have a brass or copper pot available, by all means use it, as the results obtained will be more colourful and dramatic.

1 Delegate a responsible student to manage the source of heat. If the students are very young, the teacher should do this.
2 Put the dyestuff in the enamel basin and cover it with as much water as possible. Squeeze the plant material down so it is all wet. Put the basin on the heat and bring it to a slow boil. Cook out the plant in this manner for one-half to one hour. Stir it down occasionally. (This may be done the night before, and the plant material strained off from the dye liquor. The liquor can then be poured into clean plastic jugs and carried to the classroom.)
3 Soak out the wool fibre to be dyed for at least one hour prior to dyeing in lukewarm water. (This may be white wool yarn, raw fleece, or small scraps of white wool cloth. Acrylic yarn will also take a dye with dyestuffs such as onions and marigolds.)
4 Using the strainer, strain off the dye liquor from the cooked out plant material. Discard the dyestuffs (in a strong plastic bag lined with newspaper) and pour the liquid back into the basin. This is the dyebath.

Table 6: Amount of mordant to use in workshops

weight of fibre	mordant	amount
less than 4 oz (114 g)	vinegar or urine	½ to 1 cup (120–240 ml)
less than 4 oz (114 g)	salt	¼ to ½ cup (60–120 ml)
less than 4 oz (114 g)	baking soda	1 Tbsp (15 ml)

5 Now add the wet fibre and increase the heat so that the dyebath comes to a slow simmer. (Never boil it once the fibre has been added.) The longer the yarn is in the bath, the darker and brighter the resulting colours will be.

6 Encourage the students to watch what happens as the yarn begins to take on colour. Let them decide whether or not to add a mordant. If they wish to add one, take out the fibre, add the dissolved or diluted mordant, and then put the fibre back in. (Pouring it directly on the yarn would spot it.) Liquid mordants can be added right to the bath, but powders (salt, baking soda) must first be dissolved in boiling water. Although baking soda alters colours drastically, and produces interesting results, it effervesces and care should be taken to see that it does not bubble over. (See p 24.)

Table 6 gives the amount of mordant to use for best results. You can use more than this, except with baking soda. Too much soda will cause the bath to bubble over and the yarn to feel sticky. ALL MORDANTS LISTED HERE ARE SAFE TO USE IN COOKING UTENSILS. Simply wash them out after using with warm water and soap. If urine is used, disinfect the pot by rinsing it with a solution of water and household chlorine bleach. If you have large pots and sufficient plant material to dye more than 4 ounces of fibre, simply increase the amount of mordant accordingly. Always use as much water in the dyebath as the pot can hold.

7 After dyeing is finished, more fibre can be put in the 'exhaust' bath if there is still some pigment in it.

8 Rinse all dyed fibres in warm water and then hang over a plastic rope line or lay flat to dry on garbage bags cut open and spread out on the floor. Do not hang wet fibre to dry on a painted or wooden surface. Hanging the yarn over a metal object might cause the yarn to streak where it is in contact with the metal.

Collecting

In almost every case, dyeing is more meaningful if the children themselves collect the plant material from which the dyes are made. Have students ask their parents to save onion skins. The time of year often dictates what is available. In spring, you can use alder twigs and dandelions; in early summer, wild and garden flowers; in the fall, late flowers; in the winter, bark from firewood, tea bags, or even curry powder or paprika. (The cost of these spices may be prohibitive. If using them, however, try dyeing eggs rather than yarn. The white shells take on lovely shades that are often brighter than those obtained with yarn. This is a good Easter project.) Emphasize that no living plant should be destroyed to make a dye, and make certain that anyone collecting flowers understands not to pull up the root. Although collecting urine for mordant will not appeal to every teacher, urine was, after all, a traditional mordant and is still used extensively in some cultures in other parts of the world. (In some African cultures, goat urine is used to rinse out gourds which serve as milk containers.)

Older students or those in regular art classes may enjoy using some of the chemical mordants (see p 25). However, they should familiarize themselves first with how these work and the proper procedures to follow to eliminate risk of accidents. Allow no eating, drinking, or smoking in areas where students are using chemical mordants. Utensils used with chemicals cannot be used subsequently for the preparation of food.

If dyeing is anything less than pleasurable or relaxing, it may be that the group is too large. Break the class up into units of five or six, and have them take turns dyeing on different days. Otherwise, a demonstration approach can be used. This has shortcomings with students eager to participate by actually doing the dyeing themselves.

9

Plants for dyeing

CHOOSING A DYESTUFF

The validity of using a specific plant for a dye should be considered by the dyer before – not after – the dyeing. There are many factors to consider. Certain dyestuffs are known to be easy to use. They are generally those which contain a large amount of pigment that is easy to extract. Onion skins are such a dyestuff, as are many garden flowers (calendula, marigold) and most common weeds (clover, goldenrod, lupin). If some dyestuffs are 'better' than others, it is not because they produce a more desirable colour, but because they are simply easier to process and use. Colour cannot be accurately described as 'good' or 'bad.' The desirability of any shade is wholly personal and a subjective response which varies from one person to another. When reference is hereafter made to, say, 'a good range of yellow or browns,' 'good' means that the colours are fairly fast and pleasing in their variety.

Among those factors to consider before choosing a dyestuff are the following:

1 Does the plant grow abundantly near the dyer's work premises?
2 Does the plant have unusual growth habits (eg, a high, climbing vine; thorns)?
3 Is the plant easy to collect (eg, soft, flexible stems)?
4 Is the plant large, or extremely small?
5 Is the plant easy to prepare for processing?
6 Can the pigment be extracted after several hours of soaking and one hour of cooking out?
7 Is the bath highly odiferous?
8 Does the plant contain sufficient pigment to use it for a strong bath?
9 Will a variety of mordants used with this plant produce a wide range of colours?

10 Is the dye fast to light and washing?
11 Will the same plant, collected from the same spot another year, result in a similar shade?
12 Can this plant be counted on as a reliable source of a specific colour range?
13 Is the plant rare, or an endangered species?

COLOUR

The colour any plant yields in the dyebath is affected by a variety of climatic and botanical circumstances. The geographical region within which a plant grows determines how much sun and rain it will receive. The soil in which the plant grows is an important determining factor, because the presence of certain minerals will 'show themselves' in the resulting colour when the plant is used in a dyebath. For example, soil which is unusually high in iron results in plants that give greens and khaki when they might normally be known to give yellow and yellow-greens. Plants growing in regions where there are gypsum formations will reflect this alkaline factor in the dyebath, giving strong, bright colours. Dyestuffs growing where there is fog much of the year will yield different results than the same plants growing where there is none. In some localities, individual plants become hybridized over a period of time; such idiosyncracies are important factors not only in correct plant identification but also in understanding how such seemingly trivial details can affect the colour a plant yields. Trivial to the dyer, perhaps, but never to the botanist. And once the dyer becomes experienced with identifying species, what once was trivial will become important.

Dyeing in Cape Breton is not the same as dyeing in New Zealand or California, because dyestuffs are subject to so many influencing factors which affect the colours they give. Each plant is unique. An understanding of this uniqueness will enable the dyer to feel satisfied no matter what colour does, or does not, materialize in the dyepot. Whatever the shade may be, it is special, and one of a kind. It is short-sighted for beginning or even experienced dyers to measure their colours by the results printed in a reference written thousands of miles away. The challenge of dyeing is to discover new colours of one's own, using a plant that perhaps has been overlooked by others. Therein lies the true potential for individual research in the field of plant dyeing.

THE SAMPLE BOOK

Many excellent cooks are just that because they invent and improvise, but even the best have old stand-bys or favourite recipes. It is important for dyers to keep a sample book not for other people's benefit, but for their own. The sample book is the dyer's record, or documentation, of the dyeing process used for the results obtained, samples of which should accompany each 'recipe.' A small swatch of yarn several inches long is sufficient to remind the dyer of the colour. Include one swatch for each mordant tried with each dyestuff. Give the details of where the plant was found; the date it was collected; whether it was wet or dry; and how long the processing took. Did you soak it out for several days or just an hour before cooking out? This is, in fact, the 'recipe.' Dyeing cannot be reduced to measurements alone. It is also process. But by all means, do include how much dyestuff you used for the amount of fibre dyed. Make the sample book your diary, so to speak. If it is messy and bulging at the binding, that's your concern. The sample book loses its significance if it becomes a pretty thing one carries around to show off. It should be a working reference describing what you are doing with plant dyes.

DYESTUFFS AND THEIR LOCALE

With the exception of indigo and woad, the dyestuffs listed can be found growing in eastern and central Canada and the northeastern and central United States. Some of the species are indigenous to the region. Others are aliens that have either been brought to this continent for domestic cultivation or accidently travelled here as seed in cargoes of foodstuffs or other imported goods. Some, like goldenrod and dandelion, are common over much of the continent. Many wildflowers and weeds listed grow far beyond the treeline of northern Canada and may flourish as far north as Alaska and the Yukon. This is the case with some of the lichens.

Local Department of Agriculture staff and university botanists are generally extremely helpful in assisting a dyer with plant identification. However, sadly, the relationship between the dyer and the botanist has often been unhappy. Perhaps this is natural, as plants are the botanist's professional concern. It benefits the dyer to learn conservation methods at the outset and thus not alienate professionals who might later be of assistance. Do not pick rare or endangered species. However, in order to follow this advice, one has to know plant identification. Check with local horticultural and naturalists' societies and museums in your areas to determine what plants may be scarce. For instance, I do not advocate the use of bloodroot (*Sanguinaria canadensis*) unless it has been grown by the dyer from nursery stock or purchased from a dyer's supply house. Even the latter approach is subject to criticism: these same dye houses foster the collection of these plants, although in some cases the species in question are grown commercially for the express purpose of providing dyestuffs.

HOW DYEPLANTS ARE LISTED

Dyeplants are listed alphabetically by their common names. For example, Swiss chard is listed under C and black walnut under W. The Latin name is given in addition to the common or vernacular name, as this facilitates correct identification and the exchange of information among dyers. Purslane is pussley in parts of the United States and stinking Willie refers to a variety of yellow-blooming wildflowers, as well as the domestic marigold. But pine is *Pinus* in Canada, Europe, Asia, and elsewhere. Plant identification is especially crucial when collecting mushrooms, so get into the habit and soon it will be second nature.

There are two parts of every Latin name for a plant: the first is the genus name and the second the species name. For example, tansy is *Tanacetum vulgare*. *Tanacetum* is the name of the genus the plant belongs to, and *vulgare* is the species. In some references, the Latin name is followed by a letter, or part of a name (*Tanacetum vulgare* L.). The L is the initial of the botanist who first gave the plant that classification.

Each dyeplant is further described by one of the following terms: tree; shrub; ornamental or flowering shrub; garden flower; weed; wildflower; herb; vegetable; fruit; lichen; mushroom. Weeds, wildflowers, and garden flowers are also classified as annual, biennial, and perennial as this helps the dyer to understand their habit of growth. Such information is essential to dyers who wish to start a dye garden. An annual is any plant that completes its life cycle in a single season (marigold, petunia, lamb's quarters). A biennial is a plant which produces foliage one year, and flowers and seeds the next, thereby having a two-year life cycle. Some biennials are parlsey, burdock, foxglove, and hollyhock. Perennials regrow each season from an already established root, or else reseed themselves and regenerate in that manner. Goldenrod, buttercup, delphinium, poppies, and coreopsis are common perennials. Pansies are usually thought of as annuals, but because they often reseed themselves they are treated as perennials by some gardeners. Coreopsis also reseed themselves, so the term 'annual' may be confusing to dyers new to gardening.

For the purpose of this book, a weed is simply any plant growing where one does not wish it to grow. A wildflower differs from a weed only in that it has a bloom we find attractive. 'Weed' applies to those wild plants having no conspicuous bloom or those which we consider a nuisance.

The bibliography contains a full list of all references used. Where possible, Canadian sources are preferred for Canadian dyers. Those books selected and referred to are not necessarily the only ones to consult. But some, such as R.C. Hosie's *Native Trees of Canada* and Roger Tory

Peterson's *Guide to Wildflowers*, are considered by professionals to be excellent for those new to plant identification. The photographs and drawings in both these books are unmatched in detail. This is an important factor in selecting a book for plant identification. Mason Hale's *How to Know the Lichens* is a classic reference and one which is essential for dyers using any lichens in their work.

REFERENCES FOR TREES AND SHRUBS

Brooklyn Botanic Garden Record, Plants & Gardens *Handbook on Conifers* 60, and *The Hundred Finest Trees and Shrubs for Temperate Climates* 25. Baltimore, MD: Brooklyn Botanic Garden 1969; 1957

Hosie, R.C. *Native Trees of Canada* 7th ed. Ottawa: Canadian Forestry Service, Department of the Environment 1973

Knobel, Edward *Identify Trees and Shrubs by Their Leaves: A Guide to Trees and Shrubs Native to the Northeast* New York: Dover Publications, Inc. 1972

Petrides, George A. *A Field Guide to Trees and Shrubs* Peterson Field Guide series. Boston: Houghton Mifflin Company 1972

Saunders, Gary L. *Trees of Nova Scotia: A Guide to the Native and Exotic Species* Bulletin 37. Nova Scotia: Department of Lands and Forests 1970

Sherk, Lawrence C. and Arthur R. Buckley *Ornamental Shrubs for Canada* Publication 1286. Ottawa: Research Branch, Canada Department of Agriculture 1974

REFERENCES FOR FLOWERS, WEEDS, AND HERBS

Cunningham, G.C. *Forest Flora of Canada* Bulletin 121. Ottawa: Forestry Branch, Department of Northern Affairs and National Resources 1975

Erichsen-Brown, Charlotte *Herbs in Ontario: How to Grow and Use 50 Herbs* Aurora, Ontario: Breezy Creeks Press 1975

Frankton, Clarence and Gerald A. Mulligan *Weeds of Canada* Publication 948. Ottawa: Canada Department of Agriculture 1974

– *Common and Botanical Names of Weeds in Canada* Publication 1397. Ottawa: Canada Department of Agriculture 1975

Herwig, Rob *128 Garden Plants You Can Grow* New York: Collier Books 1976

Martin, Alexander C. *Weeds* Golden Nature Guide series. New York: Golden Press 1972

Peterson, Roger Tory *A Field Guide to Wildflowers of Northeastern and North-central North America* Boston: Houghton Mifflin Company 1968

REFERENCES FOR WILD FOODS

Gibbons, Euell, *Stalking the Wild Asparagus* Field Guide Edition. New York: David McKay Company, Inc. 1973

MacLeod, Heather and Barbara MacDonald *Edible Wild Plants of Nova Scotia* Halifax: The Nova Scotia Museum 1976

Stewart, Anne Marie and Leon Kronoff *Eating from the Wild* New York: Ballantine Books 1975

Wigginton, Eliot, ed. *Foxfire 2* Garden City, NY: Anchor Books, Anchor Press/Doubleday 1973

REFERENCES FOR MUSHROOMS

Bigelow, Howard E. *Mushroom Pocket Field Guide* New York: Macmillan Publishing Company, Inc. 1974

Groves, J. Walton *Mushroom Collecting for Beginners* Publication 861, rev. Ottawa: Canada Department of Agriculture 1973

– *Edible and Poisonous Mushrooms of Canada* Publication 1112. Ottawa: Research Branch, Canada Department of Agriculture 1972

Shuttleworth, Floyd and Herbert S. Zim *Non-Flowering Plants* Golden Nature Guide series. New York: Golden Press 1967

REFERENCES FOR LICHENS

Erskine, J.S. *Common Lichens* rptd from *Journal of Education* (April 1957) Halifax: Nova Scotia Museum 1958

Hale, Mason E. *How to Know the Lichens* Dubuque, Iowa: Wm C. Brown Co. Publishers 1969

SEED CATALOGUES AND NURSERY CATALOGUES

The following are useful because many common plants are easily recognizable in their excellent photographs. Most firms listed are sources of plants, nursery stock, and seeds (Sheridan does not ship plants). Each catalogue is referred to by the abbreviation which follows the name:

- C.A. Cruickshank Ltd, 1015 Mount Pleasant Rd, Toronto, Ontario, M4P 2M1 (CK)
- Dominion Seed House, Georgetown, Ontario (DO)
- McConnell's Nursery Co. Ltd, Port Burwell, Ontario, N0J 1T0 (MC)
- Sheridan Nurseries, 700 Evans Ave, Etobicoke, Ontario M9C 1A1 (SH)
- Stokes Seeds Ltd, Box 10, St Catharines, Ontario, L2R 6R6 (ST)
- Vesey's Seeds Ltd, York, Prince Edward Island (V)

Absinthe

Weed *Artemisia absinthium*

Absinthe is also known by the vernacular names of wormwood and mugwort. It is a tall-growing perennial reaching five feet or more in height, blooming from late July through September. Although the tiny yellow-green flowers are inconspicuous, absinthe is readily identifiable by its strong aroma. Farmers dislike absinthe because cows which have fed on it produce tainted milk.

Parts used: the whole plant
Processing: as for whole plants (see p 71)
Colours obtained: (Unless otherwise stated, all colours given are those obtained using either bleached or natural white wool of various plies.) yellow-green with alum; gold with chrome; soft green with blue vitriol; khaki with iron
Fastness: excellent for all shades, but yellow-green with alum may fade in strong sunlight
How to identify: Frankton and Mulligan; Peterson, *Wildflowers;* see p 87 for other weed references. Two Peterson references are used: *Trees and Shrubs* and *Wildflowers*. *Trees and Shrubs* is written by Petrides.
Availability: found throughout Canada and the northeastern United States, growing in waste places, along roadsides, and on croplands
Special notes: absinthe may be mixed with other weeds in a single dyebath.

Acorn

nuts from oak *Quercus*, various spp.

The red oak, *Quercus borealis* (Saunders; Hosie uses *Quercus rubra;* see p 192), is the most common oak of the northeastern hardwood forests. There are several other species common in parts of Ontario and Quebec, including *Q. alba* (white oak) and *Q. meuhlenbergii* (Chinquapin oak). Acorns are the nuts, or mature fruit, of the oak. Although some oaks attain a height in excess of 100 feet (30.5 m), even much smaller trees annually yield enough acorns to provide many dyebaths. Acorns are collected when they are fully ripe, from September through the fall.

Parts used: whole nuts. For dyeing with oak bark and leaves, see pp 71–5.

Processing: as for nuts (p 75). Crush first.
Colours obtained: tan to medium brown with chrome; dark brown with iron (strong bath); golden brown with chrome and tin (all from nuts)
Fastness: excellent, but colours may darken with time (see p 40)
How to identify: Hosie; Petrides (Peterson series); Saunders. Oaks are easily identified by their large, shiny-green lobed leaves, which turn rust in the fall and often remain on the tree into winter.
Availability: Red oak occurs in pure stands in parts of Cape Breton and Queen's County, Nova Scotia (Saunders 58) but normally grows in mixed stands with aspen, birch, beech, and maple.

Nursery stock is available from SH and also from most local nurseries which carry ornamental and shade trees. Amateur gardeners are often unsuccessful in their attempts to germinate acorns for planting; also, young oaks can be difficult to transplant from the wild. Halifax weaver Florence Margeson and her family helped me dig oaks for transplanting in late October 1977, and these are now well established.

Alder

small clumped weed, tree, or shrub *Alnus rugosa*

It is no exaggeration to say that the common speckled alder is the bane of the farmer's existence. It quickly crowds out other tree species and constantly threatens to overrun cleared pastures and fields. No one will mind your collecting alder leaves, bark, or roots. Indeed, no one will care if you remove the entire clump. On the eastern coast, alders most often grow as a shrub, from 8 to 12 feet (2.4–3.6 m) in height, but can reach the size of a small tree. Alders thrive in clumps along stream beds, in windrows, and at the edges of most cleared agricultural land. Alder is useful to the dyer both as a dyestuff and a mordant (see p 27).

Parts used: leaves, bark (twigs), roots
Processing: process leaves as for fresh leaves; bark as for bark; roots as for roots. To use as a mordant, add alder twigs to the regular dyebath or cook out the twigs first and then add the strained off liquid to the dyebath.
Colours obtained: leaves give a yellow with alum; yellow-green with alum and blue vitriol; olive-green with blue vitriol; tan with iron; yellow-orange with tin. Bark and twigs give a tan to rosy brown with chrome and baking soda; dark brown with iron. Roots give a greyish-brown with chrome and iron, and variations of a greyish-charcoal shade if processed with an iron mordant in an iron pot (strong dyebath).

Fastness: excellent for all dyes from leaves, bark, and roots. Browns may darken with time.
How to identify: The yellowish male catkins form the previous autumn. The dark brown or black woody seed cones may remain on the tree all winter and serve as a useful means of identification. In the fall, alder leaves are distinctively unattractive, turning an odd bronze shade. The bark of the alder is, as the name implies, speckled in appearance and grey-brown in colour. Hosie; Saunders. Locally, ask any farmer to point out alder to you.
Availability: common throughout Canada and the northeastern and central United States. Alder often invades city suburbs as well; look for it at the edges of housing developments, baseball fields, and playgrounds.
Special notes: Once a clump of alder is established, it spreads rapidly. For this reason, it is unwise to propagate the species intentionally. There is always enough alder to serve the dyer's needs.

Amaranth

weed, annual *Amaranthus retroflexus*

The necessity for dyers to use Latin nomenclature in identifying dyeplants is readily apparent when dealing with a plant such as amaranth. Frankton and Mulligan give it three common names: green amaranth, redroot pigweed, and rough pigweed. The common name of redroot pigweed is also used by a Canadian reference devoted exclusively to nomenclature (*Common and Botanical Names of Weeds in Canada*). The issue of what to call amaranth besides amaranth is further confused when checking American dye books such as Weigle's *Natural Plant Dyeing* (p 22) and plant references, for Peterson and Gibbons use pigweed to describe lamb's quarters, which is *Chenopodium album*, quite another plant altogether. This is the kind of frustration a dyer new to plant identification encounters. To eliminate confusion entirely, just learn to identify this plant as amaranth. An annual, amaranth reaches 3 to 4 feet in height (.92–1.22 m) and grows in all Canadian provinces except Newfoundland (Frankton and Mulligan 54). Its Canadian vernacular name comes from the fact that the weed has a red root.

Parts used: whole plant. Although I have not tested the red root, it bears investigation as a potential source of red.
Processing: as for whole plants; alone or mixed with similar weeds
Colours obtained: soft yellow-green with alum; strong chartreuse with tin; brown with chrome and iron

Fastness: excellent with alum and tin mordants; fair to good with chrome
How to identify: Frankton and Mulligan; Martin; Peterson
Availability: amaranth is found in agricultural areas and waste places throughout Canada (except Newfoundland) and much of the United States.
Special notes: dyers investigating the red root are advised first to make certain no soil clings to the root after it has been collected, as this would brown the resulting bath. Also, temperatures above a simmer would be likely to encourage a brown or greyish result rather than the desired red.

Apple

fruit tree *Malus*, various spp.

There are basically three types of apple trees, the leaves and bark of which are all useful to the dyer: the wild apple or crab-apple, the ornamental or 'flowering' crab-apple, and the domestic 'eating' apple, grown for its fruit. Dyers always should first seek the wild species before attempting to collect leaves or bark from cultivated trees. Because flowering crab-apples are planted as ornamentals, collecting bark from these is inadvisable unless the owner is pruning and there are branches readily available. Domestic apple trees are often heavily pruned by orchard owners in the spring, and rural dyers should have no difficulty tracking down a good source of bark. City dyers can take advantage of the opportunity to collect apple bark by visiting suppliers of fireplace wood. Abandoned apple orchards are still to be found in many rural areas of eastern Canada and the northeastern and central United States.

Parts used: fresh leaves, bark, roots, trimmings from the fruit
Processing: process leaves as for any fresh leaves; bark and roots as for bark and roots. Trimmings from jelly or pie-making may be used, but even in a strong dyebath yield a very pale colour.
Colours obtained: from the leaves, pale yellow with alum; gold with chrome; rust with chrome and tin; strong yellow-orange with tin; soft grey with iron. From the bark or roots, yellow-tan with alum; medium rose tan with chrome; dark brown with chrome and iron; grey-brown with blue vitriol. From the skins and trimming of the fruit, in a strong bath, pale yellow with alum; soft tan with chrome
How to identify: Hosie; Saunders. Ask locally.
Availability: fruit and ornamentals are available as nursery stock from most local nurseries and MC and SH. Remains of old orchards are to be

found in most rural areas. Always ask before assuming you have the right to trespass.
Special notes: lovely shades are obtained using a dyebath of apple, pear, and cherry leaves. The barks, too, may be used in a mixed bath. Apple bark has a more pleasing aroma in the dyepot than almost any other dyestuff. If using fruit or the skins of the fruit for a bath, strain the mixture carefully through a cloth before adding the fibre to the bath in order to avoid having an 'applesauce' dyebath, which is extremely messy.

Arrowhead

aquatic wildflower *Sagittaria latifolia*

This plant has several common names, including broadleaved arrowhead, arrowleaf, and duck potatoes, the latter referring to its desirability as a wild food. Another plant with a similar shaped leaf is arrow arum, but it belongs to another genus, *Peltandra virginica*, and is more common in southern Ontario and the east-central United States. Arrowhead grows along the edges of swamps, marshes, streams, and bogs. It has green arrow-shaped leaves and a single flower stalk bearing small white flowers. If digging the tubers for dyestuff, it is wise to take along some kind of tool, such as a hiker's shovel. The tubers vary in size from ½ to 3 inches (1.3–7 cm), and collecting any number of them for a dyebath is tedious. I prefer to use the leaves of the plant for making a dye. This does not harm it, as arrowhead reproduces by tubers, as do potatoes.

Parts used: fresh leaves; tubers
Processing: leaves as for fresh leaves. The tubers are scrubbed carefully and then allowed to dry in the sun. Chop them up quite fine and cook out as you would any other dyestuff.
Colours obtained: from the leaves, a soft yellow with alum; gold with chrome; orange-rust with tin; from the tubers, old gold with chrome; rusty orange with tin
Fastness: excellent
How to identify: Gibbons; MacLeod and MacDonald; Peterson; Stewart and Kronoff. Both Peterson and Stewart and Kronoff illustrate arrowhead and arrow arum so the dyer can compare the two plants.
Availability: Arrowhead leaves can be collected from June on, but the tubers do not mature until fall. Arrowhead is found in the shallow waters at the edge of lakes, ponds, streams, swamps, and marshes.

Ash

tree, hardwood *Fraxinus americana* and *F. nigra*

Maple, birch, ash, oak, elm, linden, poplar, and beech are hardwoods. Coniferous trees such as pine, fir, and spruce are softwoods. White ash (*F. americana*) and black ash (*F. nigra*) are both commercially valuable woods. Black ash is used by native basket-makers to weave their famous splint baskets, and white ash is used to make baseball bats, hockey sticks, and bowling pins. However, neither species is common in Nova Scotia. White ash is fairly abundant in southern Ontario and southeastern Quebec, while the range of black ash extends right across Ontario and into Manitoba.

Parts used: leaves; bark (collected from deadfalls, pruned branches, or firewood)
Processing: leaves as for fresh leaves; bark as for bark and roots
Colours obtained: from leaves, soft yellow with alum; bright yellow with tin; medium dull grey with chrome and iron; beige with blue vitriol. From the bark, rose-tan with alum and chrome; brown with chrome and iron
Fastness: excellent
How to identify: Hosie; Knobel; Petrides (Peterson); Saunders
Availability: look for white ash in mixed stands with beech, birch, maple, and hemlock. Black ash prefers a wet, open locale and often grows along streams and rivers. SH is a source of nursery stock, as are some local nurseries. Small saplings can be transplanted from the wild.
Special notes: Mountain ash is listed under M because it is of another genus: *Serbus americana*. Owing to the relative scarcity of black ash, native Nova Scotian basket-makers import it from New Brunswick.

Aspen

tree, hardwood *Populus tremuloides, Populus grandidentata*

Aspens are actually poplars (see Lombardy poplar, p 196), and there are many species. The two mentioned here are fairly common in eastern Canada and the northeastern United States, namely, *P. grandidentata* (largetooth aspen) and *P. tremuloides* (trembling aspen). Poplars have grey bark that is smooth and often greenish when the tree is young. Poplar leaves flutter even in the slightest breeze, revealing their greyish-white undersides. From a distance the tree appears silvery when the leaves are 'trembling.' Once successfully identified, an aspen is easy to recognize. Lichen enthusiasts often collect the brilliant orange *Xanthoria* from its trunks.

Parts used: fresh leaves, twigs
Processing: leaves as for fresh leaves; twigs as for bark and roots
Colours obtained: from leaves, a clear yellow with alum; bright yellow-orange with tin; beige with blue vitriol; gold with chrome. From the twigs, soft grey with blue vitriol and iron
Fastness: excellent
How to identify: Hosie; Knobel; Petrides (Peterson); Saunders. Ask locally, as aspen is common in urban and rural areas. Lombardy poplar (*P. nivra*, var. *italica*) is a common urban species.
Availability: MC and SH catalogues. *P. tremuloides* extends all across Canada and down into the United States, while *P. grandidentata* occurs as far west as the Great Lakes. Poplar saplings are easily transplanted from the wild and make excellent ornamental trees for shade. Aspens are said to root successfully when fresh cuttings are stuck into moist soil (Saunders 42).
Special notes: although aspen bark is generally grey and very smooth when the tree is young, it may become quite furrowed on very old trees, but the greyish-green bark is characteristic.

Aster

flower, annual and perennial *Aster frikartii, Aster novi-belgii, Aster nemoralis*

Asters are both annuals and perennials. Some species grow very tall, while others, like *Aster novi-belgii* (Michaelmas daisy) are dwarf varieties common in most flower gardens. There are numerous domestic and wild species available to the dyer and gardener. *A. frikartii* is a purple domestic species which most closely resembles the wild asters we are accustomed to seeing in bloom during the late summer and fall. Domestic asters are available in shades ranging from white, pink, orchid, and mauve through to blue, purple, and violet. Some species are rust, red, bronze, and maroon. Wild asters tend to be pinkish-white or blue-violet. Look for wild species along roadsides and ditches, often in the company of chicory.

Parts used: fresh blooms, separated by colour or used in a combined bath
Processing: as for fresh flowers

Colours obtained: from domestic varieties, all blooms in the blue-violet range: yellow with alum; yellow-green with blue vitriol and iron; grey with chrome; bright yellow with tin. From wild varieties, all blooms pink to mauve, no stems or leaves: yellow-beige with alum; tan with chrome; greyish-green with blue vitriol and iron.
Fastness: excellent for domestic varieties; good for wild species
How to identify: Herwig; Martin; Peterson. Excellent photographs in most seed catalogues
Availability: both domestic and wild species are late bloomers; wild types may last until October on mainland Nova Scotia. Wild asters grow along roadsides and in fields, often in the company of chicory and goldenrod. As seed, from CK, DO, MC, SH, ST, and V. As bedding plants, from most nurseries
Special notes: asters, like marigolds, give excellent colours when picked for a dyebath after they have been touched by frost. Resulting shades are usually rust (with bronze and gold blooms, chrome mordant), olive-green (mixed colours, blue vitriol, and iron mordants) and khaki (blue and violet blooms, iron mordant).

Bachelor's button

flower, perennial: domestic and wild *Centaurea cyanus*

Also known as cornflower, bachelor's button is found in the wild in those locations where it has 'escaped' from cultivated gardens. Bachelor's button is a blue flower, but in the wild it may also be pink or white, according to Peterson (p 362). It grows up to two feet (.61 m) in height and blooms from July through the fall. Dyers new to plant identification may think that bachelor's button and chicory look alike, and they do. But chicory grows much taller and its unique blue flowers close up by noon on a hot day.

Parts used: fresh flowers. With wild species, the whole plant may be used.
Processing: blooms as for flowers (can be mixed with similarly coloured flowers of other genera); whole plant as for whole plants
Colours obtained: from flowers, yellow (alum) to beige; bright yellow (tin); gold to tan (chrome). From the whole plant, chartreuse (alum and tin); grey-green (blue vitriol and iron)
Fastness: excellent. Chartreuse from the whole plant bath may change to a dull yellow-green upon exposure to light over a prolonged period of time. This is also the case with the chartreuse obtained from lupin.
How to identify: Herwig; Martin; Peterson. Most seed catalogues have photographs of domestic species.

Availability: as seed, from CK, DO, ST, V; as bedding plants from local nurseries

Banana

fruit, imported *Musa*

Banana skins which have been left to ferment in water in a covered plastic container yield an excellent range of colours. Skins from mature fruit are the best to use, especially if they are mottled or spotted with black. The skins from eight or ten ripe bananas will give enough pigment to dye 4 oz (114 g) of fibre. Chop the skins up in small pieces, cover with water, and then place the covered container in a warm spot. Stir the mush every day. If the smell becomes offensive, add a dash of baking soda to the brew. Adding a little sugar also helps. When ready to use the mush, strain the liquid through a wet cloth, squeezing well to get it all. After dyeing, rinse the yarn in several baths of soapy water to remove all traces of fruit particles.

Parts used: skins from mature fruit
Processing: see above
Colours obtained: beige (alum); dark brown (blue vitriol); tan (chrome); warm gold (tin); grey-brown (iron). The brown with blue vitriol is a rather surprising result. It is probable that colours obtained will vary from one bath to another, depending upon the type of banana used, where it was grown, and how long it has been in storage.
Fastness: good
Availability: bananas are expensive. You can buy them at a special low price, mash the fruit, and then freeze it for later use. Or, start the fermentation using one or two skins and add more to these as you buy and use more fruit. Expect any bath kept longer than ten days to smell quite foul during the dyeing. If this smell lingers on the dyed fibre, rinse it in Fleecy or another fabric softener.

Barberry

ornamental and wild shrub *Berberis*, various spp.

Barberry is a traditional dyeplant used in European and Scandinavian countries as a fine source of strong yellow dye which is obtained without the use of mordants. Grae, Krochmal, Lesch, and Robertson all give recipes for barberry leaves and/or bark. However, most species of barberry in Canada have been banned, as the shrub acts as an alternate host of wheat stem rust. Frankton and Mulligan advise that the importation and propagation of deciduous barberries is now prohibited in this country. (Deciduous shrubs and trees lose their foliage each fall and regrow them the following spring, whereas evergreens or conifers retain their foliage year round.) One such species is the favourite dye source Japanese barberry, *B. thunbergii*). *B. vulgaris* (the species cited by Robertson) is also banned. Sherk and Buckley advise against planting any deciduous barberries but do provide a list of acceptable and immune evergreen species. Unfortunately, few of these will grow in eastern Canada as most are hardy only in the Niagara Peninsula area of southern Ontario. One species, *B. verruculosa*, is hardy to zone 6, which includes parts of southwestern Nova Scotia, eastern Cape Breton, and northern Prince Edward Island. (Sherk and Buckley, *Ornamental Shrubs for Canada*. Hardiness maps at the back of the book indicate what plants will grow where.) It is interesting to note that Petrides (Peterson series, p 193) considers *B. vulgaris* to be the most susceptible of all barberries to wheat stem rust. Ironically, it is that very species which occurs most frequently in the wild in this region. But dyers are urged NOT to transplant these shrubs in their own gardens. If you think you must have a barberry, try *B. verruculosa* or any other species recommended by a local nursery which specializes in ornamental shrubs. Nurseries, by law, are only allowed to offer for sale those species approved by the Canada Department of Agriculture. Dyers living in the United States will no doubt discover that barberries are not illegal in all states.

How to identify: Frankton and Mulligan; Knobel; Petrides; Sherk and Buckley

Basswood

shade tree *Tilia* spp.

Basswood is an introduced species in eastern Canada, but it grows naturally in parts of Ontario and Quebec. It is also known as linden, or lime linden. Fibres from basswood bark were used by the Indians to make a strong rope (Hosie 288). Dyers familiar with the Public Gardens of Halifax, Nova Scotia, will be interested to know that the large trees lining the park's boundaries on Sackville and South Park streets are lime lindens. When these are pruned, bark is available by asking the garden's superintendent.

Parts used: fresh leaves and bark
Processing: leaves as for leaves; bark as for bark
Colours obtained: from the leaves, strong yellow (alum); brilliant yellow (tin); gold (chrome); taupe (iron). The bark was not tested.
Fastness: excellent
How to identify: Hosie; Knobel; Petrides; Saunders
Availability: MC and SH offer nursery stock, but most local nurseries do not offer this species. Check for seedlings where basswood is planted as an ornamental or shade tree, and ask if you may have a sapling.
Special notes: Krochmal (p 95) refers to basswood by the Latin name of *Liriodendron tulipifera*, but this is quite another species of tree.

Bayberry

ornamental shrub *Myrica pennsylvanica*

This species is often confused with barberry (p 97), sweet gale (p 145), and sweet fern (p 212). However, once successfully identified, it is easy to tell which is which. *M. pennsylvanica* is a deciduous shrub, which in the wild can reach a height of 8 feet (2.44 m) and have an almost equal spread. Bayberry also has the characteristic grey waxy berries used to make bayberry candles. Gardeners are advised that the sexes are located in separate shrubs so more than one must be planted in order for berries to be produced.

Parts used: leaves, berries (Davidson, *The Dye Pot* 5, gives a recipe for obtaining blue from the grey berries of *M. cerifera*.)
Processing: leaves as for fresh leaves
Colours obtained: the leaves give a strong yellow with alum; gold with chrome; brilliant yellow with tin; greyish-green with iron. One source gives grey-green as the shade obtained using alum-mordanted fibre (Edward Worst, *Dyes and Dyeing* 28). However, I have been unsuccessful in obtaining grey-green with any mordant other than iron.
Fastness: excellent for all but the grey-green with iron, which faded somewhat in the light after several weeks' exposure
How to identify: Knobel; Petrides; Sherk and Buckley
Availability: from SH as nursery stock and some local nurseries specializing in ornamental shrubbery. Wild bayberry can be found in most wooded areas where the soil is sandy, growing amid spruce, fir, birch, and beech. Unfortunately, it often grows in locations similar to where sweet gale and sweet fern occur. This adds to the confusion when first

learning to identify each species. But both gale and fern are smaller and have highly aromatic leaves.

Bean

garden vegetable *Phaseolus*

Most home gardeners grow yellow and green beans for table eating. Usually the bean vine itself is discarded after the mature vegetable is picked. Aside from its value as compost, however, the bean vine is an excellent dyestuff. Vines from several species of bean may be combined in a single bath. Grae gives a recipe for dyeing with the beans themselves, in this case, *Phaseolus vulgaris*, or what we know as red kidney beans (p 175). The beans are cooked out first in water and this liquid then becomes the dyebath. However, in our part of North America red beans are very expensive, and it takes 8 oz (228 g) of them to make enough dye for 1 oz (28.34 g) of fibre. The resulting colour Grae gives as terracotta brown.

Parts used: bean vines of any species; liquor from red kidney beans (see above)
Processing: pull the vines from the ground and shake all soil off the root. Tear, shred, or chop the vine and leaves as finely as possible. Place in a plastic pail and cover with water. Leave in a warm location for several days, stirring the mixture at least once a day. (It will smell, so place the pail outdoors or in the garage or basement.) Cook out, and strain off the liquor. A strong dyebath is recommended for maximum colour.
Colours obtained: beige to yellow-beige with alum; tan with blue vitriol; brownish-grey with chrome; bright gold with tin; greyish-brown with iron; grey-green with blue vitriol and iron processed at a low temperature
Fastness: fair to good for all shades. The gold with tin was the most fade-resistant of the colours tested.
How to identify: There are people who would not know a bean vine unless it was pointed out to them. If you are one of these, ask neighbours for their vines or drive out to the nearest produce stand and ask to collect vines from their fields. Be honest, though, and say you don't know which is which. Otherwise you may end up half a mile from the farmer's house in the midst of a five-acre plot not knowing which are bean vines and which are squash, peas, or cucumbers. Ignorance of plants is not a crime, but the only way to learn is to have the courage to tell people that you honestly don't know. That is a positive starting point.

Availability: from your own or neighbours' gardens; from market gardens in rural areas; as seed, from DO, MC, ST, V, and most local seed suppliers
Special notes: dyers having access to the pods discarded after the harvesting of beans used for baking (cow beans, pinto beans, lima beans) might try these as a dyestuff.

Bedstraw

wildflower, weed *Galium boreale* and other spp.

Northern bedstraw, lady's bedstraw, and yellow bedstraw are the common names for various species of *Galium*. One Canadian weed reference (*Common and Botanical Names of Weeds in Canada* 16) lists five species with five different vernacular names, including those above and false cleavers and cleavers. *G. tinctorium*, listed by Leechman (p 35), is what he refers to as dyer's bedstraw, whose roots are 'the best source of red.' But *G. tinctorium* does not occur as a listing in any other botanical or dyeing reference I checked. *G. verum* is given by Davenport, Robertson, and Thurston. That is also the species referred to as the so-called cheese rennett plant (Eustella Langdon, *Pioneer Gardens at Black Creek Pioneer Village* 51). Robertson also lists *G. boreale*, which is what Peterson calls northern bedstraw. The confusing diversity of species is significant to the serious dyer because the roots of bedstraw are said to be a source of red. I have not personally tested any *Galium* species. The most descriptive explanation of how to use *Galium* is given by Robertson (p 68). The tops and leaves are also a source of yellow. In Shand's article 'Dyeing Wool in the Outer Hebrides' (p 62), she refers to the use of *G. verum* for red, and explains the discrepancies in the resulting colour as depending upon where the plant was harvested. It is interesting information and well worth the trouble to look it up.

Parts used: Leechman, p 35 (roots); Robertson, p 68 (flowers, leaves, and roots); Thurston, p 18 (roots)
Processing: see above. It may well be that *Galium* species do yield a red from the root in some climates, but not all. The Hebrides are certainly similar in climate to parts of Nova Scotia and Newfoundland, and even the soil has some of the same characteristics. Therefore, theoretically at least, bedstraw roots from species of *Galium* collected here should yield red.
Colours obtained: yellow (from flowers and leaves); red (from roots)
How to identify: Cunningham; Langdon (*Pioneer Gardens*); Peterson
Availability: Cunningham gives the locale for *G. boreale* and *G. triflorum* as Nova Scotia, Newfoundland, Quebec, and north as far as the Yukon;

west to British Columbia. Peterson lists the above species as growing throughout the northeastern United States and much of Canada.
Special notes: Galium is, in fact, a member of the madder family. Wild madder, *G. mollugo*, grows from Ontario south to Ohio and Virginia, according to Peterson (p 40).

Beech

hardwood tree *Fagus grandifolia*

Known as American beech, this native species occurs throughout the Maritime provinces, southern Quebec, and southern Ontario, as well as most of New England and the central and south-central United States. Wild food enthusiasts gather beech nuts each fall, and these are a source of food for many small rodents. The leaves of the beech were used by early settlers as a mattress filling, apparently being more buoyant than straw (Saunders, *Trees of Nova Scotia* 56).

Parts used: leaves, nuts, bark from firewood or felled trees
Processing: as for each category
Colours obtained: from the leaves, yellow (alum); rust (in a strong bath with tin); gold (chrome); tan (blue vitriol); greyish-tan (iron). I have not tested the nuts. From the bark (mature, from firewood), yellow-tan (chrome and alum); brown (chrome and tin)
Fastness: excellent for leaves; good for bark colours
How to identify: Hosie; Knobel; MacLeod and MacDonald; Petrides; Saunders; Stewart and Kronoff
Availability: beech does not occur in Newfoundland. Look for it growing among other hardwoods such as maple and birch, or scattered amid hemlock and spruce. Beech is available as nursery stock from SH but not from most local nurseries. I have tried to transplant it from the wild with some success, despite the difficulty of finding a sapling that was not a shoot from the parent tree. (Such saplings have their own roots, but grow attached to the parent tree.) However, dyers may have good results propagating beech from the nuts themselves.

Beet

vegetable *Beta vulgaris*

Among the various controversies that exist in the field of dyeing is the issue of whether or not beets give a red dye. Rather than place myself irrevocably on the affirmative or negative side of the argument, I shall instead attempt to indicate the wide diversity of opinion on this subject. The individual dyer can then decide if beets produce a fast red dye. The Krochmals (p 155), Grae (p 56), and Leechman (p 36) offer recipes for a red from beets. Grae's method is, in fact, one of steeping. Leechman's recipe to dye one pound of fibre calls for 4½ pounds of beets. The Krochmal recipe calls for 1 cup (225 ml) of alum and either fresh or canned beets. In the latter case, the sugar added to the product probably makes the resulting colour more fast than if fresh beets were used (see berries, p 75). I tried the Krochmal recipe using fresh beets and obtained a fast medium tan. However, owing to the large amount of alum used, the quality of the fibre was definitely impaired. Grae advises that her steeping technique produces a lovely red which is nonetheless fugitive. Her process involves using four or five fresh beets and 1 oz (28.34 g) of fibre. There is no cooking at all; the vegetable-water-fibre solution merely sits and soaks until the yarn has taken on the pigment. Davenport lists a brown from beet roots with an alum mordant but gives no recipe as such. Lesch gives a tan from beets, but advises they are not a good dye substance, an opinion confirmed by Davidson in her book. Adrosko, Furry and Viemont, Robertson, Thresh, and Thurston do not list beet as a dye source. However, it is safe to say that almost all beginning dyers try beets at one time or another, perhaps because they have beet-stained clothing or tablecloths and are determined that any stain so stubborn must be able to be made into a good dye. Pickled beets leave a more permanent stain than do the regular vegetables, so it may be that the addition of sugar and vinegar would affix a beet dye more readily. But investigation of the beet as a dyestuff is hardly gratifying. For the persistent, try the leaves, although the Krochmals advise that these are useless as a dyestuff.

Parts used: leaves, root of the vegetable, canned, pickled, or fresh
Processing: process leaves as for fresh leaves. Cook out the root of the vegetable (the edible beet) after chopping it up finely. Leave the skins on. The solution may be steeped rather than cooked, as the Grae recipe suggests.
Colours obtained: from the leaves, in a strong bath, yellow with alum and tin; tan with chrome and vinegar; from the root of fresh beets, tan (with alum), and red, according to sources listed above
Fastness: the shades from the leaves were moderately fast to light and washing. The tan from the root, with alum, was quite fast but the yarn was too sticky to be useful. The red Grae obtains from steeping is, according to her, fugitive.
How to identify: look at the supermarket for a red, bulbous-shaped, tuberous vegetable with large green and red-veined leaves. The sugar beet is another species.

Availability: DO, MC, ST, and V. As seed from local suppliers

Begonia

flowering bulb *Begonia*, various spp.

The begonia is a tuberous flowering bulb much prized by gardeners for its beauty and versatility. It may be grown outdoors, indoors, in window boxes, or in hanging planters. The large, velvet-looking blooms range in colour from white through to deep red, bronze, rust, rose, scarlet, and crimson. There are yellow, orchids, apricot shades, and oranges, too. Fresh blooms are far too lovely to end up in the dyepot, so wait until the flowers fade. Ask friends and neighbours who garden to save their begonia flower heads for you. They are used wilted, or after the first frost, and may or may not be separated as to colour.

Parts used: faded, wilted, or frost-bitten blooms
Processing: collect by colour range or mix shades together, cover with water in a plastic container, and add to this more flowers as they become available. Allow them to soak out for several days once you have enough, and then cook out in the same water, adding more if necessary. Strain the cooked out blooms and proceed with the dyeing.
Colours obtained: a wide range of shades can be obtained, depending upon the colour of the flowers and how they were processed. I obtained a lovely green using mixed orange, red, and rust blooms with an iron mordant, after a soaking period of one week. From frost-bitten blooms that were mixed in colour, a yellow-orange resulted with alum and tin; a gold with chrome; and a fine greenish-grey with iron.
Fastness: good, especially the yellows and golds
How to identify: Herwig; CK catalogue has many pages of excellent photographs of a wide variety of begonia species
Availability: as bulbs, from CK, DO, and MC, and some local suppliers

Birch

hardwood tree, ornamental *Betula*, various spp.

Even the most urban-oriented individual can usually recognize a birch by its white, paper-like bark. Several species are native to the northeastern part of the continent, including yellow birch (*B. allegheniensis*); paper birch (*B. papyrifera*), and grey birch (*B. populifolia*). It is interesting to note, however, that the southern limit of the range of the paper birch, also called the canoe birch, is such that in the days before the white man came to North America, some of the Indian tribes in the central United States had to come north to buy their canoes from the Algonquians living around the shores of the Great Lakes. It is tempting to remove bark from birch trees as it appears to be half-fallen off, curling outwards in small strands or coming off in large, slab-like pieces. But removing bark from a living tree may kill it if enough is taken, so get into the habit of conservation and rely on deadfalls and firewood as a source of bark. Often there will be a small pile of shredded bark at the base of a birch that has fallen off during wind storms. Collect this from a fair-sized birch stand and you will have more than enough bark. If you must resort to using fresh twigs, collect them in the early spring, from trees growing in a well established, mature stand. Remove only one from each branch. The inner bark of birch gives a stronger range of colours, but it should never be taken from living trees.

Parts used: leaves, bark
Processing: as for each category
Colours obtained: good yellows, golds, and tans from the leaves with alum, tin, and chrome; outer bark, pale yellow to soft tans with blue vitriol and chrome; inner bark, orange with tin or purplish brown with iron (see below)
Fastness: excellent for all leaf shades, good for bark colours
How to identify: Hosie; Knobel; Petrides; Saunders
Availability: as nursery stock, from MC, SH, and most local nurseries. Birch is easily transplanted from the wild.
Special notes: Robertson (p 64), Davenport (p 114), and Thurston (p 18) all list the inner bark of birch (*B. alba, B. pendula*) as a source of purplish-brown with an iron mordant. I was unable to obtain this, but did get a grey with a dusty tinge, rather like a taupe.

Blackberry

wild and cultivated fruit *Rubus allegheniensis*

The common wild blackberry is a favourite dyeplant with dyers, not so much for the colours which the berries yield as for the fine shades obtained from both the leaves and new shoots of this thorny shrub. This new growth can be used from early spring right through until fall. The purple-tinged fall leaves of the blackberry are an excellent source of grey, as are the mature 'canes' or arching branches of older bushes. Because blackberry and raspberry (see p 201) are both species of *Rubus* and often grow intertwined along the edges of pastures and fields, they can be collected interchangeably and used together in the dyepot, although there seems to be more grey-colouring pigment in the blackberry. Use the fruit for dyeing only if there is an over-abundance. Many families depend upon such berries as a food source, and if this is the case in your area, then concentrate on using the shoots, canes, and leaves of the plant.

Parts used: berries, if sufficiently abundant; new shoots, with leaves; mature canes, cut up; fall blackberry leaves
Processing: process according to type; process shoots as for whole plants. Cut up all shoots and canes first, into two- or three-inch (5–8 cm) lengths, being careful of the thorns. Use heavy garden gloves and pruning shears or a small hatchet to chop mature canes. Cutting canes or new shoots will not endanger the shrub as it reproduces readily so long as the root itself is not disturbed.
Colours obtained: from the fruit (very ripe, strong bath): pink-tan, orchid, and purple, using sugar and flour in the dyebath as suggested on page 75 (tested by Dawn MacNutt, Dartmouth, NS); fresh leaves, yellow (alum); bright gold (chrome); soft orange (tin); purple autumn leaves: beautiful grey (iron); brown (chrome); mature canes, same as for fall leaves: new shoots, yellow-green (alum and blue vitriol); warm golden brown (chrome); bright yellow-green (tin); greyish-green (iron)
Fastness: excellent for all shades
How to identify: Cunningham; Gibbons; Knobel; Petrides; Stewart and Kronoff
Availability: as nursery stock, from MC and some local nurseries. In the wild, blackberry grows throughout eastern Canada and the east-central United States. Although *R. allegheniensis* does not occur in Newfoundland, other *Rubus* species do, including red raspberry, *R. strigosus*. Wild bushes may be transplanted – new shoots, approximately 12 inches in height (30.5 cm) are easiest to dig up.
Special notes: Thurston obtains a black from new shoots with iron (p 18). I was unable to verify this, however.

Black-eyed Susan

---

wild and cultivated flower, perennial *Rudbeckia*, various spp.

Black-eyed Susan is also commonly known as coneflower, although some references distinguish between the two *Rudbeckias*. Peterson (p 112) calls black-eyed Susan *Rudbeckia hirta*, and coneflower *Rudbeckia triloba*.

Common and Botanical Names of Weeds in Canada (p 29) refers to black-eyed Susan as *R. serotina* and coneflower as *R. laciniata*. Peterson calls the latter green-headed coneflower (p 114). To further confuse the issue, another yellow flower with a brown or black velvety centre closely resembles *Rudbeckia*, and that is sneezeweed, *Helenium autumnale*. Some *Heleniums* are called sunflowers, and they tend to have light-coloured centres rather than the brown or black centre of the *Rudbeckia*. Gardeners buying the cultivated perennials for their gardens will no doubt be supplied with either *R. laciniata* or *R. hirta gloriosa*, which is known as the gloriosa daisy. Beginning dyers should not worry unduly about correctly identifying the wild species, but if you are persistent, concentrate on the difference in leaf characteristics to determine which species you have located.

Parts used: flowers, or flowers and leaves. Do not pull up and use the whole wild plant, as *Rudbeckia* is a perennial and regrows each year from the root stock.
Processing: as for fresh flowers. Cultivated and wild species may be used in a single dyebath.
Colours obtained: from wild species, *R. hirta*, flower heads and some leaves, in a medium bath: deep olive-green (iron); avocado green (blue vitriol); light greenish-yellow (alum and blue vitriol). Robertson reports obtaining a warm golden yellow using the flower petals only, with chrome (p 70).
Fastness: excellent, perhaps the most fast of all greens obtained from common flowers
How to identify: Herwig; Martin; Peterson; photographs in most seed catalogues. *Rudbeckia* species grow from 1 to 3 feet (30.5–90 cm) in height, along roadsides, in waste places, and frequently near old homesteads. They bloom from late July through September in central Nova Scotia, but will flower earlier in New England and Ontario.
Availability: as seed, from CK, DO, MC, ST, V. As bedding plants from most local nurseries specializing in perennials

Black Walnut

See walnut; also butternut.

Bloodroot

woodland flower *Sanguinaria canadensis*

One of the earliest of spring wild flowers to bloom, bloodroot is, as its name implies, an excellent source of strong red dye. It is referred to in most dyeing references, and widely used, but I cannot recommend it as a dyestuff unless it is purchased as such from a specialty supplier (see suppliers, p 228). Bloodroot occurs too sparingly in this part of Canada for dyers to collect it from the wild. There are, however, alternatives. One is to purchase it from a supplier, as suggested; the other is to order the plant from a nursery and propagate it yourself. I have done this but should, in all fairness, warn you that it is expensive and difficult to grow successfully. Only one of my several plants survived and this was after planting them exactly as recommended to me by the nursery. But do try to grow bloodroot if you are a fairly experienced gardener, and can invest in a dozen plants.

Parts used: root
Processing: the root is dug and left to dry in a warm place. Tap off all dirt. Chop it as finely as possible, or use a kitchen blender. Put the root pieces, fibre, and water to cover in the dyepot and leave to sit in a warm place for two days. Stir often. Then raise the temperature of the dyebath to a simmer and process for one to two hours.
Colours obtained: dyers who have used the commercially available bloodroot say it produces a tan-orange with alum, a tan-pink with vinegar, and a red with tin, but that the red is highly variable, ranging from rose to barn red rather than crimson or scarlet.
Fastness: reported to be fairly fast
How to identify: Peterson
Availability: as nursery stock, from CK. Their supply varies from year to year, so order it early and carefully follow all planting instructions. Bloodroot requires a moist, woodland setting and rich soil. As a dyestuff, bloodroot is also not available at all times. The price is usually high as well. If you attempt to transplant wild bloodroot plants, use great care and relocate them in exactly the same type of setting.

Blueberry

shrub, wild and cultivated *Vaccinium* and *Gaylussacia*

Because blueberries (*Vaccinium*) and huckleberries (*Gaylussacia*) often intermingle where they grow, it is appropriate to deal with them together even though they are of separate genera. Huckleberries are somewhat larger and darker coloured than blueberries, and *Gaylussacia baccata* grows from Newfoundland as far west as Saskatchewan (Cunningham

93), so it is widespread. The so-called 'low-bush blueberry' is *V. angustifolium*, while the 'high bush' varieties include *V. corymbosum*, which has scarlet leaves in the fall and red stems in winter, making it highly ornamental. Euell Gibbons (*Stalking the Wild Asparagus* 39) offers illustrations of both *Vaccinnium* and *Gaylussacia* to show the difference between the two, and Petrides illustrates a wide variety of *Vaccinium* and *Gaylussacia*, including species commonly called bilberry, box huckleberry, mountain cranberry, and farkleberry (Petrides 360). Robertson gives a recipe and refers to *Vaccinium* as bilberry (p 21). As is the case with all other berries commonly collected as a foodstuff, it is up to the individual dyer to decide if there is sufficient abundance to allow for experimentation in the dyepot. In some years the wild crop is quite scarce while in others there are blueberries and huckleberries to be had everywhere.

Parts used: very ripe fresh berries; mush remaining after jelly-making. Leaves of both *Vaccinium* and *Galussacia*
Processing: when dyeing with all berries, add sugar to the bath to improve fastness (see p 75). Vinegar may also be added, but the resulting colours may grey somewhat. Process berries as for berries and leaves as for leaves.
Colours obtained: most dyers are unaware that blueberry and huckleberry leaves are a fine source of yellow and yellow-green. Don't strip the leaves off a few plants – take some from many, over a fairly wide area. The leaves give yellow with alum, brilliant yellow with tin, and yellow-green with blue vitriol and iron. From the fruit, a few dyers succeed in obtaining pinks, orchids, and purples from blueberry and huckleberry, but I have personally been unable to get a fast shade other than a pinkish-tan or a pinkish-grey. Susanne Maclaughlan, East Gore, Nova Scotia, sent me samples of pink, purple, and blue-grey from huckleberry on white acrylic fibre without mordants. My tan was with chrome and vinegar, the grey with iron and salt. I used a medium bath, and in a strong bath did manage to obtain a slight rose shade using tin, but it was not fast to light and faded after two weeks' exposure.
Fastness: shades from the leaves are excellent in fastness; the pinks and purples seem to vary. The tan and grey I obtained were quite fast.
How to identify: Cunningham; Gibbons; MacLeod and MacDonald; Peterson; Sherk and Buckley (In *Edible Wild Plants of Nova Scotia*, MacLeod and MacDonald offer an excellent method of identifying *Vaccinium* and *Gaylussacia*: the former have many seeds and the latter [huckleberry] only ten.)
Availability: as nursery stock, from MC and many local nurseries. May be easily transplanted from the wild if relocated in an acid soil. Common in the wild, especially in burned-over areas and fields adjacent to

railroad tracks. Blueberry fields are also identifiable by the lovely aroma of sweet fern (*Comptonia peregrina*), which often grows there as well (see p 212).

Burdock

prickly weed, biennial *Arctium minus*

This annoying weed grows abundantly in most farmyards, and if used as a dye source should be collected before the burs are fully developed. It is a true biennial: the first year, only rhubarb-like leaves appear. These are woolly in appearance, as large as rhubarb leaves or even larger, but, unlike rhubarb, white underneath. The burdock roots of this first-year growth are a favourite food of wild food enthusiasts, and Stewart and Kronoff (*Eating from the Wild* 11) write that the Iroquoians used the leaves as a potherb. Gibbons says the plant has a reputation as an aphrodisiac (*Stalking the Wild Asparagus* 46), but despite such fascinating information regarding the homely burdock, dyers have generally ignored it as a source of pigment.

Parts used: the large (up to 1 foot [30.48 cm] in length) leaves from the first year's growth
Processing: shred, tear, or chop and cover with boiling water. Let this mixture sit a day or so in a warm spot, and stir occasionally. Then proceed as for leaves or whole plants.
Colours obtained: yellow (alum); tan (vinegar); yellow-green (alum and iron); strong yellow (tin)
Fastness: good
How to identify: Frankton and Mulligan; Gibbons; Martin; Peterson; Stewart and Kronoff
Availability: common everywhere in the northeast, especially around farms and along rural ditches

Burningbush

ornamental shrub *Euonymus atropurpurea*

Burningbush is also known as wahoo (Sherk and Buckley 76), and grows both in the wild and as a cultivated species valued for its purple fall foliage and red winter twigs. As with all ornamental shrubs, dyers may take advantage of spring pruning to collect bark for dyeing, or obtain it from the wild shrub.

Parts used: bark, in the form of twigs
Processing: as for bark
Colours obtained: an orange-tan with alum and tin; a pink-tan with vinegar in a strong dyebath (wild species)
Fastness: good
How to identify: Peterson; Sherk and Buckley; some nursery catalogues have photographs
Availability: as nursery stock from MC and SH. Available as stock from some local nurseries or those specializing in ornamental shrubbery

Butter-and-Eggs

wildflower, perennial *Linaria vulgaris*

The common names for *Linaria vulgaris* are many: toadflax, yellow toadflax, Dalmatian toadflax, and wild snapdragon. There is also a blue species, *L. canadensis*. But here I will concentrate on the yellow variety. I prefer the label butter-and-eggs because it so charmingly describes this very lovely wildflower regarded by Frankton and Mulligan as an ornamental introduced to this continent via Eurasia (*Weeds of Canada* 150). *Linaria* is indeed just like a small snapdragon: part of the dragon's head is the colour of butter and the other portion is the exact orange-yellow of a good country egg. I have dug up *Linaria* from the wild and transplanted it in my garden, isolated in a spot where its rapid spread will not choke out other perennials. *Linaria* is found all across Canada and ranges as far north as Fort Smith, in the Northwest Territories (ibid.), but despite that amazing range, it seems to be quite localized. Some dyers do not know it at all; many farmers are unfamiliar with it as well. I have picked it in June in Pennsylvania and New York State and in Nova Scotia in August!

Parts used: flowers, or better, the whole plant, minus the perennial root. Butter-and-eggs has a shallow root system so be careful when removing stalks not to disturb this.
Processing: as for whole plants
Colours obtained: yellow from just the flowers, but they are so small that it takes hours to collect enough. From the whole plant, yellow-green (alum and blue vitriol); chartreuse (tin); greenish-grey (iron)
How to identify: Frankton and Mulligan; Martin; Peterson. Note that Cunningham (*Forest Flora of Canada* 44) lists as 'toadflax' *Comandra richardsiana*, and Peterson (p 70) gives a *Comandra umbellata* called 'bastard toadflax.'
Availability: as described above. Look for it along roadsides and in waste places on sandy soil.

Butternut

hardwood tree *Juglans cinerea*

Also known as white walnut, the butternut is *Juglans cinerea*, whereas black walnut is *Juglans nigra*. There are few butternuts in Nova Scotia, and most of these have been planted as ornamentals. The species is common in much of New Brunswick, southern Quebec, and the Niagara Peninsula of Ontario. Although Saunders does not mention butternut in *Trees of Nova Scotia* owing to its rarity there, Hosie (p 134) writes of the 'iodine-like' yellow dye which can be made from the husks and bark. Adrosko also mentions the dyes obtained by early settlers from *J. cinerea* (*Natural Dyes and Home Dyeing* 39). The husks of butternut nuts are, like those of the black walnut, very sticky to touch. These do not need to be separated from the nuts by opening prior to dyeing. However, if the husks are split open before the dyeing, the brown from the nut may ooze into the bath and affect the strong yellow which the husk alone produces.

Parts used: fresh leaves; husks and bark
Processing: see above. For yellows from the husks and bark, leave husks whole, but chop bark, and soak out in water to cover to 48 hours. Process at a temperature below a simmer (200°F, 95°C) for yellows without brown. For browns, open the nut husks before soaking them out, and process with or without additional bark at a simmer. Use a strong bath for good browns. Process the leaves as for fresh leaves.
Colours obtained: fresh leaves were not tested, but can be presumed to yield yellows, tans, and yellow-greens, as do the leaves of the black walnut. From the husks alone (not split open prior to dyeing): strong yellow (alum); yellow-orange (chrome and tin); orange (tin); brown (iron); grey-brown (blue vitriol). From the bark alone: similar shades, but paler and slightly less fast than the colours obtained from the husks
Fastness: excellent for shades from husks; good for shades from the bark
How to identify: Hosie; Knobel; Petrides
Availability: SH offers both *J. nigra* (black walnut) and *J. regia* (Carpathian walnut). MC offers only *J. regia*. Butternut can be transplanted from the wild or propagated from the mature nuts.
Special notes: the fact that *J. cinerea* is not offered by nurseries may possibly be explained by the fact that all walnuts have roots which produce toxic substances in the soil where they are growing. Although this fact is mentioned by Hosie (p 132), Petrides says only that some species will not co-exist with the walnuts, including apple trees and tomato plants (*A Guide to Trees and Shrubs* 135). The toxin is called juglone and inhibits the

growth of grass under the trees. Petrides does mention that both butternut and walnut husks and nuts were once used to poison fish and that this is now illegal (p 135)! Perhaps dyers should, therefore, be extremely careful when dyeing with walnut to keep dye mixtures well out of the reach of children and pets.

Calendula

annual, garden flower *Calendula officinalis*

Although often called pot or Scotch marigolds, calendula and marigold do not belong to the same genus. They resemble each other in a very general way, as both are available in shades of yellow and orange. However, marigold is far more popular as a garden flower, and has a much wider range of colours. Although technically an annual, calendula freely reseeds itself and I have grown it as a perennial for a number of years. Calendula blooms well on into the fall, like marigold.

Parts used: fresh or frost-bitten blooms
Processing: as for flowers. May be used alone, or mixed in with marigolds, asters, or zinnias
Colours obtained: fresh flowers: yellow (alum); yellow-tan (vinegar); gold (salt). From frost-bitten blooms: brilliant yellow-orange (tin); brown drab (iron)
Fastness: excellent
Availability: CK, DO, MC, and ST as seed. As bedding plants from local nurseries. The plant may be identified by photographs in seed catalogues and is also listed in Herwig's book.

Carrot

wildflower and vegetable *Daucus carrota* var. *sativa*

Although botanists disagree (Frankton and Mulligan 132) as to the origins of the wild carrot, *D. carrota*, commonly known as Queen Anne's lace, it is still closely related to the garden carrot, whose orange root we eat as a cooked vegetable. Frankton and Mulligan suspect that in some instances wild carrot may, in fact, cross with the cultivated *Daucus* (ibid.). Wild carrot has a smaller root that is white to pale yellow and resembles a parsnip. It is distinctly carrot in smell, although more fibrous and woody in texture; however, MacLeod and MacDonald (*Edible Wild Plants of Nova Scotia* 128) list the root of Queen Anne's lace as poisonous, so do resist the urge to try it. The characteristic lacy bloom of Queen Anne's lace is immediately recognizable to most people with even a slight interest in wildflowers, but the tops of domestic carrots give far more interesting shades.

Parts used: domestic garden carrot tops; wild Queen Anne's lace blooms or the whole plant
Processing: chop or tear up tops and process as for leaves (a strong bath is recommended); process Queen Anne's lace as for whole plants.
Colours obtained: from tops: strong bath gave green with blue vitriol, and tin to bloom the shade. Medium bath of tops: bright yellow (alum); dark gold (blue vitriol); orange-gold (chrome); brown (iron); bright chartreuse (tin). From Queen Anne's lace, the whole plant: yellow-green (alum and tin); beige (blue vitriol); yellow-grey (iron); tan (chrome)
Fastness: the shades from domestic carrot tops are very fast but the greens may become altered after prolonged exposure. Some turn greyer, and some darker, more like an avocado.
How to identify: Frankton and Mulligan; Martin
Availability: Queen Anne's lace is widespread in Quebec and Ontario but listed as 'rare' in the Maritimes and absent from Newfoundland, Gaspé, and the prairie provinces (Frankton and Mulligan 132). It is extremely common, however, in parts of Nova Scotia such as those where I have lived in Annapolis, Halifax, and Hants counties. I have also picked it in New England and Pennsylvania.
Special notes: In his article 'Southwest Navajo Dyes' in *Natural Plant Dyeing* (p 61), Noel Bennett mentions wild carrot, but his species is *Rumex hymenoseplaus*, which is a member of the buckwheat family.

Cattail

marsh and swamp plant, perennial *Typha latifolia*

There are two common varieties of cattail, *Typha latifolia* (common cattail) and *Typha angusifolia* (narrow-leaved cattail). Both grow in marshes and swamps, valued as an ornamental because of their velvety brown flower spikes and as a nourishing food source by wildfood enthusiasts. A dye can be made from either the green, lance-like leaves, the fresh root, or the brown flower spikes. The leaves grow around the flowering stem to form a clump, and by late summer have usually reached a height of four to seven feet (1.2–2.1 m). The leaves are best gathered before the end of summer, after which time they begin to turn brown along the edges. New spring shoots are the tastiest to eat and the best source of dye because the pigment is strong and easily extracted from the new, emerging plants. The roots are more readily dug in the spring as well, provided you know the plant and can identify it without the characteristic brown flower spike to serve as a means of distinguishing cattail from other marsh grasses. However, in most locales, some dried, bleached spikes remain over the winter.

Parts used: fresh leaves, roots, flower spikes
Processing: leaves as for fresh leaves; the roots are dug and dried in a warm spot. Make certain no dirt remains on them. Chop finely, and cook out in water to cover to which salt has been added. To make a dye from the brown flowers, collect the crumbly pollen as the spike matures (usually in late summer). In early summer the flower spike has both male and female parts. The upper, male portion falls off, leaving the dense brown female flower remaining. Soak out the brown flower spikes in water to cover for several days. With both the root and flowers, a strong bath is necessary to obtain the colours listed below.
Colours obtained: from the new shoots: yellow (alum); yellow-green (alum and blue vitriol); bright yellow-gold (chrome); bright yellow (tin). From the roots: tan (vinegar) and yellow (alum and salt). From the flower spikes: beige (alum); gold (chrome); brown (iron)
Fastness: leaves, excellent; roots, good; flower spikes, good
How to identify: Gibbons; MacLeod and MacDonald; Martin; Peterson; Stewart and Kronoff. MacLeod and MacDonald point out that cattail is not bulrush, which is an entirely different species, *Scirpus acutus*.
Availability: look for cattails along the edges of swamps, marshes, and roadsides where there are ditches that retain water all summer. I have cattails growing at the edge of my hayfield, where a small swamp is developing from water which runs off the land and collects in a depression there.

Cedar

evergreen shrub or tree *Thuja occidentalis*

The eastern white cedar is supposedly called the tree of life because Jacques Cartier used a brew made from it to treat his scurvy-ridden crew, acting on the advice of the Algonquians. Only two arbor vitae species are native to Canada (Hosie 96): *T. occidentalis* (eastern white) and *T. plicata* (western red). The common name of arbor vitae is given these trees because they are not actually cedars. True cedars belong to the genus Cedrus, none of which is native to Canada (ibid.). However, the common misnomer is used here because most people in the non-scientific community do call *Thuja* 'cedar.' The tree is quite small and conical, although ornamentals may be pyramidal or globular. *T. plicata* is a tall-growing timber species occurring west of the Rocky Mountains in Canada. An evergreen, arbor vitae has flat and branched aromatic needles, and once identified, is not easily mistaken for another species.

Parts used: foliage of wild or ornamental varieties
Processing: collect foliage tips from wild trees or pruned ornamentals. Cover with water and soak out for 48 hours. Bring to a simmer and cook out for 2 hours. Strain off foliage and proceed with the dyeing.
Colours obtained: the foliage tips give excellent fast colours and the yarn retains a lovely aroma: bright yellow (alum); yellow-orange (tin); gold (chrome); tan to brown (iron)
Fastness: excellent
How to identify: Hosie; Petrides; Saunders; Sherk and Buckley. Most nursery catalogues have photographs of arbor vitae.
Availability: from MC and SH as nursery stock; available at most local nurseries. In the wild, arbor vitae occurs in parts of the Maritimes and New England, as far west as Ontario and Pennsylvania. Hosie lists it as occurring on Anticosti Island in the Gulf of St Lawrence, but not on Newfoundland (p 98).
Special notes: there is much confusion regarding what is 'cedar.' Lesch mentions cedar as a dye source but calls it juniper, and gives no Latin name (p 108). Similarly, the Krochmals (p 151) and Davidson (p 8) list red cedar as *Juniperus virginiana*, another genus.

Chamomile

annual or perennial weed *Matricaria maritima, Matricaria chamomilla, Anthemis cotula, Anthemis arvensis*

Dyers can get bogged down with nomenclature to such an extent it may ruin their pleasure in dyeing. All one really has to know is that a plant is identifiable in order to refrain from picking rare or dangerous species. Take the case of chamomile. It is an excellent dye plant, readily available, and a pest to farmers when it invades cultivated areas. But it has a plethora of names, so that if you ask someone, 'Is this chamomile?' they may truthfully say it is, although it may be any one of the four species listed above. You should pick enough of the plant you think is chamomile to prepare a dyebath, and then note the location and date of collection beside an actual pressed sample of the flower. That way you have a record. Another season you may find another chamomile, and learn the differences between it and your original sample plant. *M. maritima* is known as false or scentless chamomile; *M. chamomilla* is wild chamomile and it has a noticeable but pleasant odour; *A. cotula* is stinking mayweed, and its smell is generally considered offensive; *A. arvensis* is corn chamomile, or field chamomile (Martin 140), and it has pleasant-smelling leaves. There is one other species which is also a type of chamomile, and that is *M. matricarioides*, or pineapple weed, so named because, when bruised, its leaves smell like pineapple. Because it is so like the chamomiles, and is also a *Matricaria* species, it makes an excellent dyestuff. Also, *M. matricarioides* is available to dyers who live in the Yukon and Northwest Territories, as well as all other provinces in Canada (Frankton and Mulligan 174).

Parts used: for the smaller species, use the whole plant; just flowers or leaves may be used from species like *M. maritima*, which grow to a height of 3 or more feet (91.5 cm)
Processing: as applicable, for flowers, leaves, or whole plants
Colours obtained: strong yellow from whole plant using alum; brilliant yellow with tin; rich gold with chrome; and yellow-green with iron. With only the leaves, resulting colours are more green than yellow. With only the flowers, colours are mostly yellow and gold.
Fastness: excellent
How to identify: Frankton and Mulligan give excellent descriptions of all *Matricaria* and *Anthemis* spp. listed as dyestuffs; see also Martin; Peterson. MacLeod and MacDonald use the spelling 'chammomile,' and write that the dried flowers of *A. cotula* and *A. arvensis* make a good tea (*Edible Wild Plants of Nova Scotia* 123).
Availability: most species listed are common throughout eastern Canada and the east-central United States. As previously stated, pineapple weed (*M. matricarioides*) is listed by Frankton and Mulligan as growing in the north and northwest of Canada, but Peterson suggests a more limited range, notably southern Canada and south into the central United States (*Field Guide to the Wildflowers* 166). Look for chamomiles around farmyards, roadsides, parking lots, and playgrounds. It flowers in late June or July.
Special notes: Grae uses *A. cotula*, which she refers to as an annual having a smell she herself finds pleasant (p 80), and Thurston uses an English species, *A. tinctoria*, for bright yellows (p 20).

Chard

garden vegetable *Beta*

Swiss chard is a common garden vegetable which has a large green leaf that resembles spinach. However, chard is somewhat lighter in colour and usually lasts longer in the garden. You can either use the leaves for a dye, if you have an overabundance of the vegetable, or save the daily cooking water and make a dyebath from that. Cooking water from several vegetables may be combined for a dyebath – say, for instance, water from chard, spinach, and beet greens.

Parts used: green leaves
Processing: as for fresh leaves. A strong dyebath is recommended. Wilted or otherwise less than desirable chard can be used successfully for a dye.
Colours obtained: beige to soft yellow (vinegar); tan (iron); soft yellow-green (blue vitriol)
Fastness: good
How to identify: ask friends or relatives who garden. Chard is pictured in seed catalogues.
Availability: as seed, from DO, MC, ST, V, and most local suppliers

Cherry

fruit and ornamental tree *Prunus*, various spp.

The wild cherries which grow in this region include *P. virginiana* (chokecherry), *P. pennsylvanica* (pin or bird cherry), and *P. serotina* (wild black cherry). Of these, only wild black cherry attains tree size. Chokecherry and pin cherry are usually less than 20 feet (6.2 m) in height and chokecherry grows in a clump. Chokecherry fruit is gathered for jelly making and is supposed to have been an ingredient in pemmican, a foodstuff upon which the plains tribes depended for survival. Domestic cherry species include both those which bear edible fruit (Bing, Montmorency, Oxheart), and the ornamentals such as flowering almond (*P. triloba*). Saunders (p 240) refers to the cyanic acid present in black cherry twigs and foliage, and Hosie calls it 'prussic acid' (p 235). Apparently this substance is poisonous to some domestic animals but harmless to others, including deer. Petrides calls the chemical 'hydrocyanic acid' and writes that it is dangerous to horses and cattle, but not humans (p 235). Saunders mentions that cough remedies are made from the stewed bark of black cherry (p 64), but MacLeod and MacDonald consider the bark to be poisonous (p 128). The fruit is harmless and is used by wild food enthusiasts for everything from pie to jelly to spirits.

Parts used: fruit of wild or domestic varieties; leaves of wild or domestic varieties; barks and root of wild, pruned, or diseased ornamentals
Processing: as applicable, for berries (fruit), leaves, and bark
Colours obtained: fresh leaves, chokecherry: yellow-green with alum; beige with blue vitriol; gold with chrome; grey with iron; light gold with tin. Fruit, chokecherry: tan to grey with vinegar and salt; with alum and tin there was a slight pinkish cast to the resulting taupe. Bark, chokecherry: medium grey with iron, and purplish-grey with alum
Fastness: excellent for leaves and bark; good to fair for fruit
How to identify: Gibbons; Hosie; Knobel; Peterson; Saunders; Stewart and Kronoff
Availability: wild species border hayfields and grow amid maple, birch, and poplar. A wide variety of species are available as nursery stock from MC and SH. Most local nurseries offer fruit and ornamental stock.
Special notes: roots of the domestic variety *P. avium* are reported to dye reddish-purple by Davenport (*Your Yarn Dyeing* 115) and Leechman (*Vegetable Dyes from North American Plants* 27).

Chestnut

deciduous tree, ornamental *Castanea, Aesculus hippocastanum*

Only one species of chestnut is native to Canada, *Castanea dentata*. European sweet chestnut, *C. sativa*, is a common ornamental. The familiar horse chestnuts are not chestnuts at all but quite another genus, *Aesculus*. Indigenous chestnuts were almost wiped out at the turn of the century by a fungus disease which attacked the bark (Petrides 265), and even today chestnuts rarely grow to any size before the blight attacks them. Although the nuts of *Castanea* are small and edible, those of the horse chestnut are larger and reported to be poisonous (MacLeod and MacDonald 127). Because *C. dentata* is found only in southern Ontario and parts of the north-central United States, dyers who live in the east will have to use horse chestnuts for dyeing, unless they have access to an ornamental such as *C. sativa*.

Parts used: leaves, bark, green nut husks, nuts
Processing: treat leaves and bark accordingly; for nut hulls, see butternut, page 112; for nuts, see acorn, page 89
Colours obtained: from horse chestnut: leaves, yellow with alum; gold with chrome; grey-tan with iron. Green nut husks: yellow-green with tin; grey with iron; brown with chrome. Nuts: light tan with vinegar and alum

Fastness: excellent for leaves; good for husks and nuts
How to identify: Hosie; Knobel; Petrides; Saunders
Availability: available as nursery stock from SH, species *C. mollissima,* stated to be fairly resistant to blight. Horse chestnut is commonly planted as an ornamental in urban and rural areas. It is easily identifiable in spring with its white or pink large, upright, cone-shaped flowering heads. Look for ornamental chestnuts in public parks, and arrange to gather husks and nuts in the fall as the grounds are readied for winter. Have bark saved for you if any pruning is done.
Special notes: horse chestnuts can be propagated from the nuts, but young trees grown in this manner seem to need protection from the hot sun in summer and the wind in winter. Plant them in a protected area or use a mulch.

Chicory

perennial wildflower *Cichorium intybus*

Also known as blue sailors, coffee-weed, and blue daisy, chicory is an easily identifiable wildflower because few others that have blue flowers attain as great a height (up to 5 ft or 1.5 m) and bloom from late July through the summer. The flowers may be blue, blue-white, or blue with a pinkish cast, depending upon the soil in which they are growing. Used as a coffee substitute, chicory root when roasted smells and looks much like ground coffee beans. It makes an excellent range of colours, as do the leaves and whole plant. The long, slender taproot, however, is often difficult to pull from the ground. Take a shovel when collecting it. Wild food enthusiasts eat the young leaves, which are similar in appearance to dandelion leaves.

Parts used: flowers, leaves, whole plant, roasted root
Processing: process according to type. To process the roots, dig enough to make a medium bath. Wash these well and allow them to dry in the sun for a day or so. Then cook in a slow oven (250°F, 122°C) for half an hour, turning the roots over occasionally. Increase the heat slightly and roast until the roots are brown and smell like coffee. This will probably take an hour or more, depending upon the type of heat used. When the roots are done, remove them from the oven and let them cool. Then crush with a mortar and pestle or grind in a blender until fine. Four ounces (114 g) of ground chicory root will dye 8 oz (228 g) of fibre a medium brown. The bath has a lovely aroma.

Colours obtained: whole plants: yellow-green (blue vitriol); yellow-tan (alum); green (iron); tan (chrome). Roots: light brown (alum); medium dark brown (chrome in a medium bath); deep brown (iron); khaki (blue vitriol bloomed in tin)
Fastness: excellent for whole plant shades and all root colours
How to identify: Frankton and Mulligan; Gibbons; MacLeod and MacDonald; Peterson; Stewart and Kronoff
Availability: chicory is very common, growing mainly along roadsides and in ditches. It is often found in fields that are not regularly hayed. I have found chicory in city parking lots and near playgrounds, but rural roadsides seem to be a favourite habitat.
Special notes: one interesting identifying characteristic of chicory is that its blooms close on a hot summer day by early afternoon. The flowers remain open when it is cloudy or raining. Chicory root is easily dug on the day following a rainstorm.

Chive

perennial herb *Allium schoenoprasm*

Chives belong to the same genus as onion and garlic; there are numerous wild and domestic varieties. In *Herbs in Ontario: How to Grow and Use 50 Herbs*, Charlotte Erichsen-Brown mentions *A. tuberosum*, garlic chive, which tastes of garlic and grows higher than *A. schoenoprasm* (p 7). Chives are easy to grow, and once established will provide the dyer with enough for cooking and dyeing. The seeds from the mature, purplish flower heads can be dispersed for reseeding, or the chive clumps dug up and divided into smaller clumps.

Parts used: the green foliage (like green onion leaves only shorter and much thinner)
Processing: chop finely or tear apart with the hands. (Do not use an iron-bladed knife when chopping or this will result in dull colours.) Soak out in water to cover for a day or so, and then proceed to cook out.
Colours obtained: pale yellow to medium yellow (alum); soft yellow-green (blue vitriol)
Fastness: excellent
How to identify: Erichsen-Brown; Peterson
Availability: as seed from DO, MC, ST, V. Available from local seed suppliers. Potted chives are frequently sold in supermarkets. I have been unsuccessful in keeping these alive indoors in winter, but have suc-

ceeded with chives from my own garden. Erichsen-Brown gives detailed instructions on potting chives for winter use.

Chokecherry

Prunus virginiana

See cherry, page 118.

Cinquefoil

shrub, perennial flower *Potentilla* spp.

Cinquefoil occurs domestically as a small shrub with bright yellow flowers (*P. fruticosa, P. parvifolia*), a garden perennial (*P. aurea*), and in the wild as a wildflower (eg, *P. simplex, P. canadensis*). Only the garden shrub was tested. There is also a variety which has a red-purple bloom (*P. palustris*) and is known as marsh cinquefoil. Davenport (p 117) says a red can be obtained from the root of this species, although I have not tested it myself.

Parts used: blooms from shrubs or wildflowers, whole plant (wild), root of *P. palustris* (see above)
Processing: according to type. No instructions are given for processing the root by Davenport.
Colours obtained: soft yellow with alum (flowers, domestic shrub); wild species not tested. Red may be available from the root.
Fastness: excellent
How to identify: Cunningham; Frankton and Mulligan; Martin; Peterson
Availability: as shrubs and garden flowers, from SH. Cunningham gives the habitat of *P. palustris* and *P. fruticosa* as Newfoundland, Labrador, and Nova Scotia west to British Columbia and the Yukon (*Forest Flora of Canada* 63–4).

Clay

Common red clay, and indeed, any other type of earth can be used as both a dye and a mordant (see pp 27, 41). Make up the following solution and then immerse the wetted fibre in it:

2 gallons of water worked into 2 gallons of clay or
9 litres of water worked into 9 litres of clay

This concoction may or may not be heated, but in either case move the fibre around often and turn it over from time to time. Usually, heating the mixture will give a stronger colour in a shorter time than if the mixture is unheated. The fibre may be left in the clay and water solution for a week or more. After the desired colour has been obtained, rinse the fibre carefully in cool baths and dry it in the shade. The resulting colours will depend upon (1) the clay and its mineral components, (2) the type of water used, (3) the type of fibre dyed. Clays with noticeable amounts of iron in them, for instance, produce very interesting colours. Adding vinegar and/or salt to the water and clay may act to set the colour and make it more permanent. Clay dyeing in an iron or copper pot produces fascinating colours but it is a messy business. Beginning dyers are urged to experiment with this unusual process which is, technically at least, not a plant dye. Good red and brown clays can be found throughout the Maritimes and coastal United States, and inland along riverbeds and lake shores. Prince Edward Island is often thought by tourists to be entirely made of clay when they disembark from a ferry boat on a rainy day! Staining and dyeing with clay is an ancient tradition, one which deserves much more attention.

Special notes: Lillias Mitchell describes a black dye made from lake mud at Inishnee, Ireland (p 23).

Clematis

flowering vine, wildflower *Clematis* spp.

The most common form of clematis is a climbing, flowering vine which produces beautiful flowers in shades of white, yellow, mauve, blue, and purple. In the wild, *C. virginiana*, or virgin's bower, is a climbing wildflower with small white blooms. *C. recta*, var. *mandshurica*, is a short garden flower that produces white flowers early in the season.

Parts used: flowers, leaves, and vines (after autumn pruning). Many species of clematis require cutting back after the period of bloom is over. These vines and leaves may then be used for a dye.
Processing: according to type. Process vines and leaves as whole plants.
Colours obtained: from purple domestic vine, faded blooms: yellow-green with alum; green with iron; chartreuse with tin. From vines and leaves: grey-green with iron; brown with chrome; tan with blue vitriol

Fastness: excellent for vines and leaves, good for blooms
How to identify: Cunningham; Martin; Peterson. For domestic species, see MC and SH catalogues for photographs
Availability: C. virginiana grows in Nova Scotia, southern Quebec, and southern Manitoba, and south into the United States. Domestic vines and flowers are available as nursery stock from MC and SH and some local nurseries.

Clover

wildflower, perennial *Trifolium pratense, Melilotus alba*

The clover is white sweet clover, *Melilotus alba,* while the common pink to purple-headed clover which grows almost everywhere on this continent is red clover, *Trifolium pratense*. Peterson lists almost twenty clovers, but those presented here are the ones I have worked with in dyeing. White sweet clover is distinctive because it grows to a great height, often reaching 6 or more feet (1.82 m). It grows along roadsides and throughout rural areas, where its tall, sweet-smelling white blooms flourish from June through to October. Red clover (which may have pink, red, rose, or purple blooms) is quite different in appearance from white sweet clover. It is short and easily identified because *T. pratense* usually occurs in large patches in hayfields and vacant land in urban areas. An entire field of red clover in bloom in June is a sight to behold. Not only does it smell wonderful, it tastes just as good, and I often grab handfuls of the blooms and eat them. Watch out for bees and wasps when collecting clovers.

Parts used: either species, blooms, blooms and leaves, whole plants
Processing: according to type
Colours obtained: white sweet clover (flowers and leaves): medium yellow (alum); gold (chrome); orange (tin); tan (iron); soft green (blue vitriol). Red clover (flower heads): strong yellow-green (alum); brilliant chartreuse (tin); soft green (blue vitriol). Clovers processed after the frost give olive-green and khaki shades if the whole plant is used.
Fastness: excellent for both *Trifolium* and *Melilotus*
How to identify: MacLeod and MacDonald; Martin; Peterson
Availability: look for white sweet clover along ditches and roadsides, or near abandoned farm buildings and at the edges of fields. Look for red clover in hayfields, parks, and vacant city lots. Both grow throughout Canada and the United States. There are many other species of both *Tri-*

folium and *Melilotus*, and all are useful as dyes. The tiny pink lawn clover is also suitable for dyeing, but is tedious to pick.
Special notes: soaking out either of the clovers in water to cover for ten days or so will produce an interesting range of colours, but the smell is hardly as fragrant as the fresh clover dyebath.

Coffee

ground bean *Coffea arabica*

Leftover coffee grounds make an excellent dye if mordants are used to help make the colour fast to light and washing. Coffee is now so costly that it seems wise to use what normally would be thrown away. A dyebath may be made from (1) fresh, unpercolated coffee grounds (expensive!), (2) used coffee grounds, (3) powdered instant coffee, (4) leftover coffee (liquid) collected over a period of time (without milk added; sugar is all right). The best results, in the sense of being the most fast, are obtained from fresh ground coffee which has not been percolated. But this is costly. The next most efficient method is to use coffee grounds processed in a drip or filter-type pot. This coffee, which does not 'boil,' produces grounds that still have a lot of pigment in them. Instant coffee is expensive to use as it is weaker than non-instant grounds. All colours described here are from grounds from a drip pot. Store the grounds in the refrigerator in a covered jar until enough have been collected. The cheapest coffee available to the dyer (and tea, for that matter) is what is left over in cups following a church supper or some other such event. People who do not add milk to their coffee, but drink it black or just with sugar, do you, the dyer, a great favour.

Parts used: coffee grounds, used in a drip pot
Processing: when enough coffee grounds have been collected (two 1-lb, or 453-g size coffee cans full), add them to some water to which you have added a pinch of salt and some eggshells. This amount will dye 8 oz (226.5 g) of fibre. The shells add an alkaline factor and also clarify the bath. Then add the wetted fibre to the coffee mixture and heat to a simmer. Process for one to three hours, depending upon the colour desired. Once the fibre has been dyed, rinsed, and dried, a good shaking should remove all the grounds from the skein. Some dyers prefer to cook out the grounds for an hour or more and then strain off the liquid before the dyeing begins. Fibre may be left in a coffee (or tea) bath for one or two days for darker shades.

Colours obtained: a warm tan with vinegar (using amount of coffee and fibre as specified above); a medium brown with chrome; grey-tan with iron. Using twice as much coffee, or half as much fibre, will result in darker shades.
Fastness: fibre left to soak in a strong coffee bath will be more fast than fibre processed in a weak or medium bath. None is totally fast to fading in the light, but fibre in a strong bath left to soak two days was very fast to washing when tested.
Availability: look for free sources of coffee grounds and tea bags as well. Visit organizational suppers, institutions, cafeterias, restaurants – wherever coffee and tea are used in large amounts. They may be combined in a single bath, giving colours that are more tan than brown.
Special notes: to dye cotton with coffee, use a tannin mordant (from sumac twigs or leaves. See p 40).

Coltsfoot

perennial, wildflower *Tussilago farfara*

Coltsfoot is one of the earliest of all spring wildflowers to bloom, appearing even before the dandelion. Its interesting name comes from its appearance: the stem is encased in many scales and a little imagination transforms this into the furry leg of a young foal. MacLeod and MacDonald (p 70) provide another common name for coltsfoot which is even more charming: son-before-the-father. This refers to the fact that the flower is not accompanied by a leaf. After the bloom dies down, the leaves appear and last all summer long. To my eye, they resemble the leaves of wild rhubarb, only those of the coltsfoot are naturally smaller and felt-like on the underside of the leaf. Coltsfoot is a small plant, with stems only a few inches high (7.5–15 cm) and flowers that are smaller than those of the dandelion.

Parts used: flowers (they are so small that collecting them is tedious); leaves
Processing: flowers as for flowers and leaves as for leaves
Colours obtained: colours from the flowers are much the same as those obtained from dandelion blooms (see p 134). The leaves give yellow with alum; grey with blue vitriol; tan with chrome; taupe with iron and yellow-green with tin.
Fastness: excellent
How to identify: MacLeod and MacDonald; Peterson. First locate the flowers, and you can then identify the leaves and collect them as they appear later on.

Availability: coltsfoot is a common plant in Nova Scotia, and I have seen it in Maine and New Hampshire as well. Peterson gives the habitat as far west as Minnesota, east to Nova Scotia, and south into Ohio, Pennsylvania, and New Jersey. Look for it along roadsides, growing often within inches of the pavement. In central Nova Scotia the flowers appear by mid-April and earlier when there is no snow. It could be expected to bloom in the east central United States at least a month earlier.

Copper Penny Blue

Although copper penny blue is a mineral dye, this recipe is included because the ingredients are readily available and it is easy to concoct. The pennies can be used once the dyeing is over by simply rinsing them off; they are not harmed in any way. Copper penny blue was also included on the basis of its originality: the fact that a dyer read of the method, improvised, and came up with her own 'recipe' is a lesson for us all, especially those of us who rely on the inventiveness of others! The recipe was given to me and devised by Dawn MacNutt, a Dartmouth, Nova Scotia dyer and weaver who originated a number of highly interesting recipes. Those few changes I have made are noted below. Mrs MacNutt found the idea in Oliver N. Wells's book, *Modern and Primitive Salish Weaving*.

Ingredients: 4 ounces (114 g) of fibre
⅓ cup (70 ml) of ammonia (liquid variety)
⅔ cup (140 ml) of water
50 pennies in 2 gallons (9.09 l) of water
Procedure: mix the ammonia and two-thirds cup of water together. Put the copper pennies in an enamel pot and pour the water and ammonia over them. Then add the 2 gallons of water and enter the wetted fibre. Keep this mixture COVERED as the fumes are toxic. Allow the pennies to steep in the water and ammonia bath for one to two weeks, depending upon the desired colour. The mixture is not cooked. With fleece, Mrs MacNutt obtained lovely soft shades of blue and slate grey after eight days. She recommends adding more pennies after several days if the colour seems too grey rather than blue. Fibre left in longer will dye a darker colour. The bath will also dye cotton.

When I experimented with this recipe, I used white wool yarn and found it necessary to turn the fibre over several times a day to prevent streaking. Probably, dyeing unspun fleece is best suited to this recipe, because if there is some streaking it will not matter once the fibre is spun into yarn. I used 200 pennies for 8 oz (226.5 g) of fibre, 3 gallons (13.5 l)

of water, and 1 cup (240 ml) of ammonia. The resulting shade was a soft slate-blue reminiscent of the copper-sheathed roofs of old buildings. A delicate colour, it was slightly uneven but beautiful when woven into a twill sample with off-white and soft grey yarns of similar weight. Mrs MacNutt recommends using the recipe as a top dye over fibre originally dyed in a strong yellow bath, to produce good greens.
Special notes: As ammonia is a strong alkaline substance it is normally unwise to use too much of it when dyeing fibre. But since this recipe is not subjected to any heat, the use of the amount of ammonia given should not impair the quality of the fibre dyed. It is interesting to note that Davidson (p 10) gives a similar recipe, for green, in her book using chrome-mordanted yarn with iron and blue vitriol. The bath is heated and ammonia is added as well.

Coreopsis

garden flower, some annual, some perennial *Coreopsis tinctoria*

The wild variety, *Coreopsis lanceolata,* grows west and south of the Ontario/Manitoba border in Canada and has an all-yellow flowering head, whereas the escaped and domestic garden species, *C. tinctoria,* has a brownish centre rather like a black-eyed Susan. One nursery (Sheridan) offers the wild variety, *C. lanceolata,* but most offer *C. tinctoria* for home gardeners. As a perennial, coreopsis blooms well into the fall; some species are treated like annuals which reseed themselves. Another name for *C. tinctoria* is 'tick-seed' (*Common and Botanical Names of Weeds in Canada* 11).

Parts used: flowers, blooms, and stems after frost
Processing: as for flowers
Colours obtained: fresh flowers: bright yellow (alum); bright orange-yellow (tin); rusty orange (chrome). Flowers collected after frost: dull gold (alum); dull rust (chrome and tin); brown (chrome and iron). For an olive-green, use frost-bitten flower heads, leaves, and stems with an iron mordant.
Fastness: excellent for all shades
How to identify: Peterson; photographs in most nursery catalogues
Availability: CK, DO, and SH. I found coreopsis difficult to grow from seed, but easily raised when purchased from nurseries as bedding plants. I treat mine as perennials and allow some seeds to develop, which I scatter amid the mature clump.
Special notes: Although Thresh (p 17) and Hasel ('Coreopsis for Reds on Cotton and Wool,' *Natural Plant Dyeing* 33) obtain strong oranges and

orange-reds from coreopsis, Hasel's results were obtained using Saskatchewan plants and the Thresh results were obtained in California. It is possible that the climate in both these regions contributes to the red-producing pigment obviously available from coreopsis. Robertson's results (p 50) are similar to my own, although I do not know whether or not she tested this plant in England. Hasel writes that the addition of potash contributes to the red obtained from coreopsis, and this bears investigation.

Corn

garden vegetable *Zea*

Any part of the corn plant can be used for dyeing: the stalk, the leaves, the husks which cover the ear, or the cobs after the kernels have been removed. Farmers who grow cattle corn as fodder are often willing to give some stalks away, and the husks from two dozen ears will provide sufficient dye for 4 oz (114 g) of fibre. Corn is not a strong dyestuff, but nevertheless an interesting one to use. The colour obtained from the cobs is rather bland, a pale beige, so dyers may wish to concentrate on the rest of the plant. In the 'Art in New Brunswick' chapter of *A Heritage of Canadian Handicrafts*, Huia Ryder describes a fascinating method of rug hooking which involves the use of corn, or maize (p 237). Apparently the maize was dyed green and then, while still hot, hooked into rugs. No mention is made of the source of the green pigment.

Parts used: all parts of the plant
Processing: shred, tear, or chop all parts used. If dyeing with cobs, break them in two and soak out in water to cover for several days.
Colours obtained: husks and leaves, strong bath: soft yellow with alum (ironically, exactly the colour described as 'maize'!); tan with chrome; greyish-tan with iron; soft yellow-green with blue vitriol bloomed in tin. The cobs produced a pale beige with vinegar and chrome.
Fastness: good; fair
How to identify: if you don't know what a corn stalk looks like, growing in the field, then ask. It's the only way to find out.
Availability: as seed from DO, MC, ST, and V. From local seed suppliers. In rural areas, farmers who produce corn to sell at produce stands will usually give you permission to collect stalks after the vegetable has been harvested. Arrange a mutually agreeable time and take large bags with you.
Special notes: Grae (p 177) gives a recipe using purple 'Indian' corn but I have not tried this. In Martha Stearns's article 'Family Dyeing in Colonial

New England' (*Dye Plants and Dyeing* 81), there is an interesting reference to red obtained from using red (purple) corn. The Brooklyn Botanic Gardens dyers were unable to duplicate the results but said that red corn husks would dye wool a dark grey.

Cranberry

cultivated and wild fruit *Vaccinium macrocarpon, Vaccinium oxycoccus*

Although cranberry belongs to the same genus as blueberry (see p 108), it is quite a different-looking berry to those who know and use it as a food. For one thing, it's red, while blueberries are blue. Interestingly enough, the Krochmals consider it the same berry (p 129). *V. macrocarpon* is called large cranberry and *V. oxycoccus*, small cranberry. Both are evergreen plants which produce berries quite late in the fall. Cranberries are so rich a source of vitamin C that it seems wasteful to use them for dyeing. A bath can be made from fresh or frozen berries, but I only use for dyeing the residue left over when making cranberry juice to drink. The use of cranberry as a dye source is limited, as the colours produced are not all that interesting, nor are they fast (Furry and Viemont 8).

Parts used: the berry or its juice
Processing: as for berries
Colours obtained: from the residue left over from making juice, to which an additional 2 Tbsp (30 ml) of sugar (see p 75) was added: beige (with vinegar); pinkish-tan (alum and tin); taupe (iron); chrome produced a grey
Fastness: fair for all but grey, which was quite fast
How to identify: Gibbons; MacLeod and MacDonald; Petrides; Sherk and Buckley
Availability: cranberry grows in boggy areas from Newfoundland south and west to Minnesota, and according to Petrides, as far south as Arkansas for *V. macrocarpon*.
Special notes: it is entirely possible that cranberries may produce a pink on wool if the fibre is merely steeped in a berry solution and not cooked. However, the resulting colour, like that from beets, would not be fast to light and washing, so the fibre would be limited in its use.

Cucumber

garden vegetable *Cucumis*

Although many dyers have tried using tomato vines as a dyestuff (p 215), few seem to have tried cucumber vines. These are treated just like bean vines (p 100) and give interesting colours. Cucumber vines grow prolifically and most home gardeners throw them away after the vegetables are harvested, unless they have a compost heap. The vines are tough, though, and do benefit from a long soaking out in water to cover prior to the dyeing. This is also true of squash vines.

Parts used: vine and leaves from cucumber
Processing: see above
Colours obtained: warm yellow-beige with vinegar; pale yellow with alum; gold with chrome; bright yellow with tin; grey-tan with iron; tan with blue vitriol
Fastness: good
How to identify: cucumber vines look like squash and pumpkin vines to the uninitiated, so if a neighbour invites you to pick all you want, admit that you don't know which is which rather than pull up all their squash vines and ruin that crop.
Availability: as seed from DO, MC, ST, and V; as seed from local suppliers. Farmers who grow pickling cukes for produce stands will allow you to collect vines as long as you go with suitable containers and don't allow pets or children to romp at large through the rest of their planted acreage.

Daffodil

flowering bulb *Narcissus*, various spp.

Daffodils, narcissus, and jonquils belong to the same genus of bulbous plants which highlight the home garden in early spring. Most daffodils are yellow, but some, like those offered by specialty companies, are apricot, salmon pink, and rose-pink. (See Cruickshank's fall catalogue for dozens of exotic varieties.) The more usual colours are white, creamy white, pale lemon, soft yellow, bright yellow, and yellow with reddish-edged centres. The blooms are used when faded and dyers can mix the colours depending upon what is available. Although daffodils and, indeed, all narcissi, were once planted in formal garden arrangements, they are now often seen growing in more natural surroundings, planted amid trees and in wooded areas. The bulbs multiply each year, so even beginning gardeners who start off with a mere dozen will eventually have a substantial number of clumps. Blooms occur as early as March in some areas, and usually continue into late May, depending upon the weather.

Parts used: faded blooms from any narcissi species, mixed colours
Processing: as for fresh flowers
Colours obtained: from mixed daffodil, jonquil, and narcissus blooms, in white and yellow (soaked out for three days): soft yellow with alum; medium gold with alum and chrome; bright yellow with tin; old gold with chrome and iron; grey with iron
Fastness: excellent
How to identify: CK and MC fall catalogues have excellent photographs. Herwig; Peterson (related wild species)
Availability: CK and MC as bulbs. Available at local suppliers, including department stores. Bulbs are planted in the fall for spring blooming.

Dahlia

garden flower *Dahlia* spp.

Most dyers would agree that the colours obtained from dahlia baths are uniformly glorious and interesting. Most use large dahlia blooms, but the Krochmals give a recipe for the roots (p 196), presumably referring to the tuberous bulb. Dahlia bulbs are dug up each fall and replanted in the spring, so using the tuber for a dyebath would be a shame unless it were diseased or otherwise unsuitable for replanting. Because they often grow so tall and have such large blooms, dahlias are very showy flowers. They last well into the fall, although not starting to flower until late summer. The frost-bitten blooms give interesting colours as do fresh or faded blooms. Some varieties are small, with powder-puff type blooms; others like the cactus dahlia have enormous shaggy heads that look like a lion's mane. Dahlias range in colour from shades of orange, red, scarlet, and maroon through to rust, magenta, and purple. Although the colour of the flower used does not indicate the colour of the dyebath, dahlia baths seem to give the truest shades when bloom colours are not mixed. Orange and red blooms may be mixed, and also maroon and purple. Combining a light dahlia bloom such as salmon with a dark shade often results in dull colours that belie the true potential of using the dahlia as a dyestuff.

Parts used: fresh or faded blooms
Processing: as for flowers
Colours obtained: from faded, fuchsia-coloured dahlia blooms: strong

yellow-green (baking soda); yellow (alum); dark gold (blue vitriol); green (iron); red and orange blooms: bright yellow-orange (tin); rust (chrome and tin); magenta and purple heads: grey-green (iron); green (blue vitriol bloomed in tin); brown (chrome and iron)
Fastness: excellent, but the greys and greens may change in hue slightly
How to identify: Herwig; photographs in seed catalogues. Available as seeds and tubers, from CK (both); DO (seed); MC (tubers); ST (seed), and V (seed). Dahlias are available from most local nurseries as tubers.
Special notes: Furry and Viemont (p 18) and Lesch (p 41) think dahlias are unsuitable as a dye for cotton and linen. Dartmouth dyer Dawn MacNutt reports obtaining a strong orange from faded orange and orange-red dahlia blooms using a unique method. She allowed the flowers to remain in a brown paper bag for several days, until they were a 'sodden, steamy mess.' The resulting colour was worth it! The mordant used was tin.

Daisy

wildflower *Chrysanthemum leucanthemum*

The daisy is probably one of the most easily recognized of all wildflowers. Almost everyone knows this plant and the seemingly universal 'He loves me, he loves me not' verse which accompanies the removal of the daisy's petals, one by one. Daisies are a regular component of hay, growing as they do in hayfields along with buttercups, red clover, and vetch. They invade the city, taking root in parks, playgrounds, parking lots, and between cracks in concrete sidewalks. It is tedious to pick and use only the flower for a dyebath, so I much prefer collecting the flower, stem, and leaves.

Parts used: whole plant
Processing: as for whole plants. Daisies mixed in with buttercups and other wildflowers produce good yellow-green baths.
Colours obtained: soft yellow-green (alum); bright yellow-green (tin); beige (vinegar); chartreuse (baking soda); gold (chrome); dull pale green (iron)
Fastness: excellent
How to identify: Frankton and Mulligan; Martin; Peterson
Availability: common throughout most of Canada and the United States. According to Frankton and Mulligan, however, daisy is fairly rare in Saskatchewan and that portion of Alberta south of the Peace River (Frankton and Mulligan 176).

Dandelion

weed, wildflower *Taraxacum officinale, Leontodon autumnalis*

There are two things that must be said about using the dandelion as a dyestuff. First, even the most urbanized would-be dyer recognizes the plant. Second, the widespread belief that the roots yield a magenta or red dye drives dyers to examine this plant over and over again. I found myself trying to get red from dandelion root so often it became rather boring. My results did not once approach anything near a red. However, in the interest of dyeing and because many internationally recognized authorities DO get a red from the root, it benefits all of us to at least consider the evidence. Davenport (p 115) obtained a dull yellow-brown from the root but listed it as the traditional source of magenta used in the Highlands. The Krochmals obtained a purple from the roots, using alum (p 145). Grae and Robertson do not list dandelion as a dyestuff and Lesch gives a recipe for only the flowers (p 41). In Winnifred Shand's 'Dyeing Wool in the Outer Hebrides,' she writes that the whole plant is used to obtain a magenta (*Dye Plants and Dyeing* 64). Leechman gives a red-violet recipe from dandelion root using 2 lbs of root to 1 lb of fibre, or a strong bath (p 43). In all cases, *T. officinale* is the species used. *L. autumnalis* is a fall dandelion, somewhat taller than *T. officinale*. I tested the roots of both species without obtaining anything but yellow, gold, or brown.

Parts used: leaves, blooms (separately or together), roots
Processing: as for leaves, flowers, or whole plants. Dig the roots, wash them off well, and dry. Chop fine and cook out after soaking them in water to cover for 24 hours.
Colours obtained: flowers only: yellow (alum); bright gold (tin); dark gold (chrome); tan (blue vitriol); soft beige-grey (iron). Leaves: yellow-green (alum); chartreuse (tin); warm tan (iron); gold (chrome); greyish-green (blue vitriol). Roots: tan (vinegar and salt); gold (alum); gold-brown (baking soda); grey (iron); bright gold (tin); brown (chrome)
Fastness: good
How to identify: Frankton and Mulligan; Gibbons; MacLeod and MacDonald; Martin; Peterson; Stewart and Kronoff
Availability: widespread throughout most of Canada and the United States. In urban areas, dandelion grows on lawns and most people will allow you to dig up as many roots as you wish in order for them to be free of what is considered to be the bane of the lawn-owner's existence.
Special notes: although I was unable to obtain anything close to a red from dandelion roots, at one point when mordanting a root bath with

baking soda the bath did turn a noticeable strong green. However, this colour quickly vanished and the addition of blue vitriol and iron did not bring it back. I would appreciate hearing from dyers who have succeeded in getting the elusive red or magenta.

Delphinium

garden flower, perennial *Delphinium* spp.

The delphinium is a favourite perennial which, like the lupin, has a flower spike composed of many blossoms on the end of a long stem that may grow five or six feet (1.5–1.8 m) high. Although blue and purple are the preferred shades, delphiniums also may be white, pink, or mauve. Some varieties bloom twice in a season, and mine often last on until November. The blooms are just as useful as a dye when they are faded.

Parts used: flower spikes
Processing: as for flowers. Delphinium blooms may be mixed in with other similarly coloured flowers for a single bath.
Colours obtained: from purple flowers: bright yellow (tin); soft green (blue vitriol); pinkish-tan (iron)
Fastness: good
How to identify: Herwig; photographs in most seed catalogues
Availability: as seed, from CK, DO, MC, ST, and V; as bedding plants from most local nurseries.
Special notes: The Krochmals (p 137) obtain a blue from delphinium with alum, while Davenport lists the result as a blue-grey (p 125). Alicia Marr, who lives in Sackville, Halifax County, NS, showed me a fine true blue she got from blue delphinium, but I was unable to get these results, possibly because of a difference in the species of plant or the soil.

Dill

herb, annual *Anethum graveolens*

A herb used as a pickling spice, dill makes an interesting dyebath provided it is available in the dyer's or a friend's garden. It is too expensive to be worthwhile if you have to pay for dill because the colours obtained from it are pleasant but by no means particularly outstanding. Dill smells delightful, but must be used in a strong dyebath to produce any colour other than a soft beige.

Parts used: dill plant, flower head, and stalk
Processing: chop, tear, or shred the plant and cover it with water. Allow this mixture to soak out for two or three days. Stir it occasionally and let it sit in a warm place.
Colours obtained: soft beige (vinegar, medium bath); strong bath: soft yellow (alum); strong yellow-gold (tin); medium dark gold (chrome); soft grey-green (blue vitriol, saddened in iron)
Fastness: good
How to identify: Erichsen-Brown; pictured in some seed catalogues. Look for fresh dill at produce stands and supermarkets from August on.
Availability: as seed from DO, MC, ST, and V; from local seed suppliers

Dock

weed, perennial *Rumex crispus*

Rumex crispus is also known as curled dock and yellow or sour dock. There are other common varieties of dock in this region, including *R. obtusifolius*, with which *R. crispus* often hybridizes (Frankton and Mulligan 32). Dock is a tall weed, often reaching 3 or more feet (.92 m) in height. The edges of its long, broad leaves are wavy and tiny green flowers occur in clusters along the stem and branches. In the late summer these flowers turn to reddish-brown seeds which give dock a distinctive appearance. Wild food enthusiasts enjoy the young leaves as a boiled green and the Iroquois are said to have eaten the brown seeds as well (Stewart and Kronoff 16). The root of *R. crispus* is said to yield a black used in the Hebrides (Shand 65), and Davenport mentions the black results along with a yellow-gold from dock roots (p 115). Grae gives a recipe for roots of *R. obtusifolius* processed in an iron pot, with nails, for a dark green or brown (p 84).

Parts used: whole plant, leaves, young plant; roots, seeds, and stems from mature plant
Processing: leaves and stalks from the young plant (still green) are processed as for whole plants; the roots are processed as for roots; seeds, stems, and leaves from the red-brown mature plant benefit from a long soaking out in order to soften them.
Colours obtained: fresh leaves, stalks, or whole plant: yellow (alum); bright yellow-green (tin and blue vitriol); gold (chrome); greyish-green (blue vitriol saddened with iron); roots: yellow-gold (chrome); tan (vinegar); grey-green (iron and blue vitriol); mature reddish-brown whole plants, including seeds: warm beige (alum); medium brown (blue

vitriol); medium gold (chrome); dark brown (iron); bright yellow-gold (tin)
Fastness: excellent for all shades
How to identify: Frankton and Mulligan; Gibbons; MacLeod and MacDonald; Martin; Peterson; Stewart and Kronoff
Availability: dock grows in waste places, fields, along roadsides, and in pastures. Look for it near new building sites in rural and suburban areas. I had one solitary dock plant in my backyard which provided enough dyestuff for several ounces of fibre. Once dock turns colour it is easily recognized. The reddish-brown tall stem keeps its clusters of heart-shaped seeds well into the early winter.

Dogwood

ornamental trees and shrubs *Cornus florida, Cornus alternifolia*

There are numerous dogwoods, and *Cornus florida* is one which is tree-sized, although it occurs in Canada only in southern Ontario. *C. alternifolia* grows throughout eastern Canada and the northeastern United States, reaching a maximum height of 25 feet according to Hosie (p 298) and 40 feet according to Petrides (p 76). *C. stolonifera* is the low red-osier dogwood and it, like the others, has red bark which makes this species so readily identifiable in winter.

Parts used: fresh leaves, fall leaves, red bark, or roots
Processing: according to type
Colours obtained: fresh leaves: yellow (alum); gold (chrome); beige-tan (vinegar); autumn leaves: pinkish-beige (alum, vinegar); brown (iron and chrome): roots: yellow-gold (alum); warm tan (chrome). See special notes for information on obtaining red from *Cornus* roots.
Fastness: good for all colours
How to identify: Cunningham; Hosie; Roland and Smith; Sherk and Buckley; Petrides
Availability: MC and SH offer nursery stock: some local nurseries offer those varieties hardy in this region. *C. alternifolia* grows wild. Look for its bright red twigs and branches in winter. Dogwood usually grows amid willows, alders, and poplars, often in ditches and along roadsides.
Special notes: both Davidson (p 9) and the Krochmals (p 162) report obtaining a red from the root of *C. florida*, a species I did not test. The Krochmal recipe requires no mordant. Dyers who live west of the Quebec-Ontario border or in the east-central United States may be able to grow *C. florida*. *C. stolonifera* is easy to transplant from the wild and the winter red bark brightens up the garden.

Dulse

seaweed *Rhodymenia palmata*

Dulse and many other types of kelp and seaweed are excellent dyestuffs provided the dyer is aware of the outrageous smell such a bath will produce. No other dyebath was as distressing to me, or my family, as that made from mixed seaweeds! Perhaps dyeing with dulse, then, is best done over a driftwood fire at the beach, or if you live in an urban area cook it up over the barbeque. It is possible that I happened to select particularly smelly seaweeds, but be forewarned of the possibility of driving out other members of the household. Dulse is a marvellous food, extremely high in protein and other valuable nutrients (MacLeod and MacDonald 100).

Parts used: use all the dulse, kelp, or seaweed you gather that is free from attached barnacles and, sadly, human litter
Processing: chop, tear, or shred the dulse and soak it in salt water to cover for several days. Note: salt water is hard on your dyepots, so use regular water if you prefer, or soak the mixture in plastic pails. The resulting shades will probably be different from those listed here if fresh water is used.
Colours obtained: with dulse, in a strong bath: medium warm tan (vinegar); taupe (baking soda); soft warm beige (alum); grey-tan (iron)
Fastness: excellent, if you can stand the smell long enough to process the bath for two hours
How to identify: MacLeod and MacDonald; Shuttleworth and Zim (*Non-Flowering Plants*, Golden Guide series, contains basic information on all non-flowering plants including seaweeds, algae, mushrooms, lichens, and ferns); Stewart and Kronoff. Various types of seaweeds abound along the coasts of the Atlantic provinces, New England, and the shores of the Great Lakes. They are algaes, attached to rocks which are only fully exposed when the tide is out. This limits the time when you can collect dulse, and many a novice knows at least one good tale about how they or someone else was once stranded while collecting dulse, rocks, or shells. Dulse is dark purple, blackish-purple, or red-purple in colour. It has veinless 'hands with fingers' which are anywhere from a few inches to a foot in length (6–30.48 cm) that are without float bladders, those small air pockets which characterize many seaweeds. Ask locally where to go for dulse. The shores of the Bay of Fundy are exceptionally good areas for picking dulse, although other types of kelp can be found throughout the region.

Special notes: Lesch offers a seaweed recipe, but she does not specify the type except to say it was collected in Maine (p 111). Davenport writes that dulse (*R. palmata*) collected in the spring gives a medium brown (p 116).

Elderberry

cultivated and wild shrub *Sambucus canadensis*

A shrubby plant known for its purple-red berries, elderberry is a favourite of home wine-makers, and the fruit also makes excellent jelly. One opinion of the desirability of elderberry as a dyestuff comes to us from a book entitled *Annapolis Valley Saga* (p 111). A certain Mrs Dick has a chamberpot of urine on her stove, heating it for a dye she is making 'out of elderberries and otherwise worthless ingredients'! Mrs Dick forgets her chore and the pot boils dry, leaving what is described as a 'ripe' smell. Dyers have varying degrees of success using elderberries. Grae has a recipe for blue and blue-grey from the fruit but states these shades are not entirely fast to light (p 136). Robertson gives recipes for violet, blue, and reddish-violet but does not mention how fast these colours from the fruit are expected to be. A local weaver, Susanne MacLaughlan of East Gore, NS, was successful in dyeing white acrylic fibre with elderberries. She sent me samples of a mauve, blue-grey, and a true purple. As long as these were kept in a box they did not fade (mordants unknown).

Parts used: leaves, fresh fruit
Processing: according to type
Colours obtained: as described above. The leaves give a good range of yellows, tans, golds, and browns. From the berries, or fruit, I obtained a soft pinkish colour with alum; a pinkish-tan with chrome; and a blue with a purple tinge with iron. These shades were not fast to light after one week's exposure.
Fastness: excellent for colours from the leaves; poor for colours from the fruit
How to identify: Cunningham; Gibbons; Hosie; Petrides; Stewart and Kronoff
Availability: as nursery stock, from MC and SH; available from most local nurseries. In the wild, *S. canadensis* (common elderberry, Canadian elderberry) grows from Nova Scotia as far west as Manitoba, but *S. pubens* (red elderberry) has a wider range, including Newfoundland. Look for elderberry shrubs bordering cleared fields in rural areas, where they grow amid birch, alder, blackberry, and raspberry.

Special notes: Petrides advises that all parts of *S. canadensis* are reported to contain hydrocyanic acid, like the wild cherry (see p 118). However, he notes that the fruits are widely used by humans without any apparent ill effects (p 48).

Elm

ornamental, hardwood *Ulmus americana*

Elm is listed as a dyestuff not because it produces unique colours, but rather because large numbers of these stately trees are being destroyed, thus making leaves and especially bark readily available. Unfortunately, the rapid spread of Dutch elm disease has meant that some elms from almost every city, town, and village in eastern Canada have been removed. Most of this wood is simply carted off and dumped for later burning. You can usually have all the bark you need by simply asking for it.

Parts used: leaves, bark
Processing: according to type
Colours obtained: leaves: yellow (alum); tan (vinegar); yellow-green (baking soda); gold (chrome); bark: medium grey-brown (chrome); dark grey-brown (iron); warm tan (salt and alum); beige-tan (vinegar)
Fastness: excellent
How to identify: Hosie; Knobel; Petrides; Saunders. Wherever crews are seen removing very tall hardwood trees, stop and ask if they are elms.
Availability: some resistant species are being offered by nurseries. SH offers Siberian elm (*U. pumila*), but advises it cannot ship this tree beyond Ontario.

Evening Primrose

wildflower, biennial *Oenothera biennis*

The evening primrose is so named because its yellow petals open towards evening and close during the heat of the day. A similar plant, *O. perennis*, is known as sundrops, which is rather ironic. Needless to say, it stays open during the day and is generally a shorter plant than *O. biennis*, having smaller blooms as well. Evening primrose is common along roadsides, where it is easily recognizable by its thick, reddish stem. It blooms from June onwards.

Parts used: blooms, or the whole plant
Processing: according to that part used. There are few flowers, so using the whole plant makes collection easier.
Colours obtained: flowers: yellow (alum); brilliant yellow (tin); bright gold (chrome): whole plant: yellow-green (alum and blue vitriol); greyish-green (iron); chartreuse (alum bloomed in tin); tan (chrome)
Fastness: good for blooms; excellent for whole plant
How to identify: Frankton and Mulligan; Martin; Peterson
Availability: look for evening primroses along roadsides and in ditches and other waste places. It occurs in all provinces of Canada. SH offers a domestic variety, *O. tetragona*, which has attractive lemon-yellow flowers.

Fern

non-flowering plants, embryophytes *Pteridium aquilinium, Matteucia struthiopteris*, and others

Ferns are vascular plants that reproduce by means of spores, as do mushrooms. There are so many genera, species, and types that it would be impossible to list them all. For the dyer's purposes all types of ferns can be successfully combined in the dyebath. The advanced dyer will probably wish to identify the genus and species used for future reference. There are many excellent sources of information and those recommended (see how to identify) are inexpensive and readily available. The most commonly recognized of all ferns are probably the well-known bracken (*Pteridium aquilinium*) and the source of fiddleheads, the ostrich fern (*Matteucia struthiopteris*). All ferns produce fiddleheads, which are the tightly curled immature fronds, but the ostrich fiddlehead is the one most commonly eaten. (Dr Paul Newberne, MIT, *Journal of the National Cancer Institute*, vol. 53, no. 3, discusses ferns as carcinogens. *M. struthiopteris* is safe, he found, but *P. aquilinium* caused test rats to develop tumors.) Ferns flourish from the Arctic to southern climates and number thousands of species.

Parts used: the leafy fronds at various stages of maturity
Processing: as for any other leaf. Older fronds seem to yield different shades so try ferns at various times of the year.
Colours obtained: from *M. struthiopteris*, collected in July: soft lime-green (alum and blue vitriol); yellow-green (alum, bloomed in tin); bright green (alum, blue vitriol, bloomed in tin). From same species, picked in September: gold (chrome); soft gold (alum); bright gold (alum and tin); brown (chrome and iron)

Fastness: good to excellent; some variance
How to identify: Cunningham shows two dozen varieties; Shuttleworth and Zim picture four dozen types; Stewart and Kronoff show common species such as *Pteridium* and *Matteuccia* by silhouette and offer visual comparisons.
Availability: ferns are everywhere. Some types which are especially attractive are offered as garden plants by CK, MC, and SH, including the fiddlehead, *M. struthiopteris*.

Fir

coniferous softwood tree *Abies balsamea*

To most beginning dyers, fir and spruce look alike. Both are coniferous trees, but upon closer inspection you will notice that the fir has flat needles, less dense branching, and is a shiny, bright green in colour. *Abies balsamea*, or the balsam fir, is one of eastern Canada's main exports to the Christmas tree market of the northeastern United States. Dyers are urged not to damage living specimens of this, or any other tree as the spruce budworm infestation has reduced the number of healthy conifers. Budworm affects fir as well as spruce. Use the tips of the branches of your Christmas tree, or someone else's. If you must collect living specimens, go to an area where there is a stand of fir and remove a small tree that is misshapen or crowded. Another method of collecting fir (or spruce, pine, and hemlock) is to carefully prune off the tips of branches, taking only a few from each of many trees. City dyers can ask the sanitation department's permission to follow their special 'Christmas tree pick-up' vehicle and collect as much fir and spruce as they need in that way.

Parts used: tips of branches from mature or young trees; cones
Processing: there is little point in removing the needles from coniferous branches before using them for a dye. It is simpler to tear up the branch tips or larger branches so they are in small pieces that will fit into the dyepot. Cover with water and soak out 24 to 48 hours, pushing the dyestuff down every once in a while. Cook out the branches and strain off the liquor, which then becomes the dyebath. Process the cones as you would nuts (see p 75).
Colours obtained: read about tannin baths, page 64. Fresh fir tips from mature branches (the tip is the very end of the bough, 2 to 4 inches or 5 to 10 cm in length): gold (alum and vinegar); tan (blue vitriol); warm brown (chrome); bright gold (tin). Cones: tan to medium brown (alum

and chrome); grey (blue vitriol and iron); beige (alum and vinegar); warm tan (baking soda)
Fastness: excellent. Browns may darken if processed at a temperature above a bare simmer.
How to identify: Hosie; Knobel; Petrides; Saunders
Availability: SH offers ornamentals. Fir is easily transplanted from the wild but take care to select trees from stands not affected by budworm. Those relocated in early spring seem to do better than trees transplanted later in the season. Most local nurseries offer *A. balsamea.*
Special notes: I find that fibre processed with cones from any coniferous tree retains a somewhat 'gummy' feel which is quite unattractive. Such yarn should not be used for clothing, where this quality would make the garment unpleasant to wear.

Forsythia

flowering shrub *Forsythia,* various spp.

Forsythia is probably the shrub which blooms first in spring in this region (Zone 5b on hardiness maps. Seed and nursery catalogues have hardiness maps which indicate the 'zone' for each part of the country, and in some cases, the continent. Plants are then keyed to this zoning. Those hardy in this region are described as 'hardy to zone 5,' or 'hardy zones 3 to 7.') Almost everyone recognizes its bright yellow flowers, which appear before the leaves, so eager are we to see something so splendidly alive after the long winter! The blooms may last almost a month, depending where the shrub grows and whether or not there is a late storm. Some forsythia bloom as early as March in Ontario and the southern part of New England, but in Nova Scotia most do not flower until late April or May. The flowers may be picked for use as a dyestuff after they have faded. This does not harm the shrub or deprive the dyer of its loveliness during blooming.

Parts used: flowers; or use stems and other cuttings after pruning
Processing: as for flowers. If you do not have enough, combine them with daffodils or yellow tulips.
Colours obtained: yellow (alum); bright yellow (tin); bright gold (chrome bloomed in tin). With daffodils: dull yellow (alum); deep yellow (tin); warm tan (chrome and iron)
Fastness: excellent
How to identify: Sherk and Buckley; Petrides

Availability: from MC and SH as nursery stock. Available from most local nurseries

Fungi

Polyporaceae family: non-flowering plants

The classification of various fungi constitutes an entire branch of botany. The study of *Polyporaceae* is highly specialized, so for the dyer's purposes, I shall simplify the matter as follows. Gill fungi are herein considered 'mushrooms' and bracket fungi, *Polyporaceae*, or pore fungi. *Polyporaceae* often take the form of brackets or 'shelves' which attach themselves to the trunks of trees. (Lenzites are similar to polypores in that they may grow as shelves or brackets.) The pore fungi, when examined under a microscope, appear to have pinholes on their undersides. Sometimes these are visible to the naked eye in larger specimens. Most of the bracket fungi are distinctive in appearance and have a corky texture that gives them the look of wood or leather. Within *Polyporaceae* there are many genera and species, but perhaps the most easily recognized is the so-called 'artist's fungus,' *Ganoderma*, several species. Summer camp craft programs often involve collecting these and drawing on the soft underside. Where a sharp tool pricks the white or buff-coloured surface, a dark brown line appears. This is quite permanent and makes for an interesting objet d'art. Other common shelf or bracket fungi are sulphur shelf (*Lateiporus sulphureus*) and hen of the woods (*Polypilus frondosus*). Unlike mushrooms, pore fungi do not appear and disappear with the seasons. Once established, the fruiting bodies produce new growth each season and this remains year round.

Parts used: the whole fungus. Fungi are fascinating primitive plants, and should not be collected as if they are a scourge to mankind. They have a definite place in the overall botanical scheme.
Processing: many fungi are extremely tough; however, they must be chopped up to release their pigment in the dyebath. If the bracket is extremely hard, soak it in water to cover for one to three days, or until you can easily dent the surface with your nail. Most bracket or shelf fungi are light in weight, so you may have to put a brick on top of them. If you use a rock to weigh the fungi down, be aware that the mineral content of that rock in the water may affect the colour of the resulting dyebath. If the fungi does not soften after it is soaked, try pouring boiling water over it, or heat it up in the dyepot whole, just as it is. Let it simmer

for an hour, cool it, and then try to cut it up. Save this water as the dyebath, and add the cut-up fungi to it.
Colours obtained: yellow, beige, tan, greyish-yellow, light brown, gold – most fungi give these shades. I obtained a dark brown from *Ganoderma* which was chopped up and soaked out for two weeks. The smell was unpleasant, but the fibre dyed an interesting shade. However, the colour faded after several weeks' exposure to light. All types of fungi are interesting to work with and warrant further investigation as potential dyestuffs.
Fastness: excellent for all but dark brown
How to identify: Groves; Shuttleworth and Zim. Most mushroom references treat some of the *Polyporaceae* as well.
Availability: common throughout

Gale

woody shrub *Myrica gale*

Read about sweet fern (*Comptonia peregrina*), page 212.

Two plants which are frequently wrongly identified by dyers using them are sweet gale (*Myrica gale*) and sweet fern (*Comptonia peregrina*). The confusion stems from the similarity in name and appearance. Further, both *M. gale* and *C. peregrina* have strong-smelling leaves which aid in identification. Sweet fern is not listed under 'fern,' but under 'S.' Botanically, *C. peregrina* is not at all similar to the ferns, but the foliage is quite fern-like in appearance. Successful identification of both *M. gale* and *C. peregrina* depends upon two factors: the shape of the leaf and the habitat. Please look closely at those references listed in 'How to Identify.' All offer illustrated comparisons of the leaves of both species, and this is extremely helpful, especially when you are holding the actual samples in your hand. Plant identification is guessing unless you actually collect a specimen.

M. gale is also known as bog myrtle, and belongs to the bayberry family, most members of which have highly aromatic leaves and twigs. The leaves of sweet gale are shaped like the end of an oar or a paddle; they are toothed only at the tip, which is blunt in shape. The shrub is rough and woody in appearance and attains a height of four or five feet (1.2–1.5 m). It can be found along the edges of streams and rivers, near swampy land, and where the soil is sandy and moist. (A.E. Roland and E.C. Smith, *The Flora of Nova Scotia* 332. This reference, although technical, is extremely comprehensive in scope and invaluable to the serious dyer as a plant reference. It has excellent illustrations and habitat maps for species in NS, NB, PEI, and Anticosti Island.)

Parts used: fresh leaves and twigs; young or mature leaves, with or without twigs
Processing: tear or shred leaves and twigs. Soak them out in water to cover for two or three days, stirring the mixture down occasionally.
Colours obtained: the colours obtained from *M. gale* and *C. peregrina* are very similar, except that tin with the leaves of sweet fern will give a fine orange. New leaves, without twigs: yellow (alum); brilliant yellow (tin); strong gold (chrome). New leaves with twigs: yellow-green (alum); yellow-grey (iron); brown (chrome). Mature leaves, with twigs: strong brown (chrome and iron); olive-green (blue vitriol, saddened in iron); warm tan (alum). Mature leaves without twigs give good greens with alum, blue vitriol, iron, and tin.
Fastness: good, but the greens that are soft or pale may grey slightly with time
How to identify: Cunningham, Petrides, and Roland and Smith all have comparative illustrations of sweet gale and sweet fern. See also Sherk and Buckley.
Availability: some *Myrica* species are available from MC and SH
Special notes: M. gale is more commonly used for dyeing in some countries than it is in Canada. Robertson (p 36) gives a recipe for 'bog myrtle,' as does Dagmar Lunde, in *Dye Plants and Dyeing* ('Several Colour Tones from One Dye Bath' 69).

Geranium

house and garden flower *Pelargonium* spp.

Although red was once the most usual colour for geraniums, they are also available in shades of white, pink, rose, scarlet, and red-pink. The blooms contain a large amount of pigment. Rather than use them fresh, collect geranium flowers as they fade, and store them in some water in a covered container (preferably plastic). Many home gardeners have masses of geraniums in flower-beds and window-boxes. Call a few neighbours and ask if you can stop by once a week to remove their faded blooms, or have them save these for you.

Parts used: faded flower heads; heads with some leaves added to the bath
Processing: as for flowers. Remember that a long soak in water prior to dyeing will give deeper, stronger colours.
Colours obtained: blooms, mixed colours, no leaves: green (baking soda); light green (blue vitriol); tan (iron); bright green (tin, alum); grey

(chrome, iron). Blooms, mixed colours, with leaves: olive-green (iron); strong medium grey (alum, iron); khaki to brown (chrome); greenish-grey (blue vitriol). No fresh blooms were tested.
Fastness: excellent. All shades had an interesting and attractive depth of colour.
How to identify: Herwig; photographs in most seed catalogues
Availability: as seed from CK, DO, MC, ST, and V. As bedding plants from most local nurseries. Geraniums are annuals, but those planted outdoors in summer may be wintered inside as houseplants.
Special notes: Grae gives a recipe using red geranium leaves (*Pelargonium hortorum*) which calls for 8 oz of leaves for 1 oz of alum-mordanted cotton. The resulting colour is described as purple or grey (p 122).

Gladiolus

flowering bulb *Gladiolus* spp.

Among summer flowering bulbs, the gladiolus is the most spectacular and the most popular. Its tall flower spikes of white, yellow, ivory, cream, salmon, pink, mauve, magenta, red, and even green (the MC nursery catalogue offers a green specimen called 'Lucky Shamrock') are showy and dramatic. It seems inconceivable to consider using these blooms fresh for a dyebath! Collect the faded flowers from people who specialize in growing these fine plants. Work out a mutually agreeable arrangement for picking up the blooms weekly. Many fussy gardeners will insist on their doing the picking themselves. Fine. Let them. After all, it's not your garden.

Parts used: faded blooms (no stalks)
Processing: as for flowers. Light colours are obtained by keeping the shades separate – white, cream, ivory, yellow, pale mauve. Dark shades result from the longest period of soaking, so keep that to a minimum for soft colours. Shred or tear the blooms before putting them in water.
Colours obtained: from light flowers: yellow (alum and vinegar); beige (blue vitriol); soft yellow-gold (chrome); bright yellow (tin); beige-grey (iron). Dark shades (reds, purples): green (iron and blue vitriol); bright yellow-green (alum bloomed in tin)
Fastness: good for dark shades; excellent for light ones
How to identify: most seed catalogues have many colour photographs of 'glads'
Availability: bulbs from CK, DO, and MC; most local suppliers sell bulbs

Goldenrod

perennial weed, wildflower *Solidago juncea* and other spp.

The novice dyer usually tries onion skins first, followed by marigolds or goldenrod. There are several reasons for the popularity of goldenrod as a dyestuff: it is readily available; people consider the plant a nuisance; it provides a wide range of colours; and these shades are fast. Goldenrod is a prolific plant which spreads by means of seeds and rootstock. Contrary to the popular myth, some authorities think *Solidago* plays no significant role in contributing to hay fever (Frankton and Mulligan 160). As both Frankton and Mulligan and Roland and Smith (p 656) agree that there are almost 100 (or more) species of *Solidago*, no attempt will be made here to differentiate among these. Advanced dyers, however, will probably wish to identify their species collected, and for this purpose the Roland and Smith book (*The Flora of Nova Scotia* 656–65) is recommended (see p 145). *S. juncea*, a species commonly known as early goldenrod, is extremely common in some counties of Nova Scotia, but rare in the southwest portion of the province. Conversely, there are species there that are rare where *S. juncea* flourishes. *S. juncea* is plume-like in appearance, and has bright yellow rays. It blooms in late July, whereas many other goldenrods are not in flower until August. Not all *Solidagos* look alike; some are plume-like, while others are flat-topped, elm-shaped, or tall, slender wands.

Parts used: flowering tops and whole plant, at any stage of maturity
Processing: no matter how you handle goldenrod in the dyepot, it will always reward you by producing something. Process as for flowers or whole plants. It is also useful as a dye for topping greens and blues.
Colours obtained: it is difficult to even partly indicate the range of colours available from goldenrod. It may be used to dye white, grey, and brown fibre, producing fascinating results on the darker yarns. Flowers used without any leaves produce yellows, golds, and tans. Slightly immature blooms produce yellower shades than those which are fully open. Using the whole plant gives yellow-greens and greens with mordants such as blue vitriol and iron. A good avocado to olive-green can be obtained using all the plant except the bloom, with blue vitriol and iron as mordants. Goldenrod processed in copper and brass pots produces wonderful shades of khaki and green. Soft yellows and yellow-golds may be had by using mature flowers only, and keeping the temperature of the bath below the simmer (see p 65). For browns, use mature blooms and leaves with chrome and blue vitriol. Tin produces bright, sharp golds and often a bronze unequalled in clarity. Experiment! Few dyestuffs are as available

and as prolific as goldenrod. This is an excellent plant for beginners to use, and is well suited for demonstration purposes because it never fails.
Fastness: excellent
How to identify: Cunningham; Frankton and Mulligan; Martin; Peterson; Roland and Smith. The latter gives the range of species in the maritime provinces, while Peterson indicates the range for species in Canada and the United States. Frankton and Mulligan do not list *Solidago* as occurring west of Saskatchewan (p 160). Some dyers new to plant identification may confuse goldenrod and tansy (see p 214). Tansy has a dense, flat-topped flower round in shape and medium to dark gold in colour. Although it is as tall-growing as goldenrod, the latter is far bushier in overall appearance and much more common. Also, goldenrod is usually in bloom later in the season than tansy, and lacks its highly characteristic odour.
Special notes: indigo-dyed blues are often topped with goldenrod yellows to make green. Goldenrod is excellent as a dye for wood, leather, and shells. The pigment is so strong that even a very weak bath will give the dyer results.

Grass

lawn grass various spp.

I once lived where there was an acre of lawn to be mowed at least once a week and the resulting grass clippings went either to the compost heap or my dyepot. Collect grass for use as a dyestuff right after it has been cut. Make certain that the clippings are free from oil and grease. Discard grass which is compacted into a mass.

Parts used: freshly cut grass clippings
Processing: soak out the clippings in water to cover for several hours before making the dye. Use enough to make a strong dyebath.
Colours obtained: clear strong yellow (alum); brilliant yellow (tin); gold (chrome); tan (blue vitriol); soft grey (iron). Alum and tin give a yellow-green.
Fastness: excellent
Availability: Lawns at cemeteries and parks are mowed often, usually by machines that have attached bags. Ask the person mowing to let you empty these.

Hawthorn

tree-shrub *Crataegus* spp.

Although they are not common, hawthorns are easy to identify because they have thorns and bright red fruits that look like small apples. *Crataegus macrosperma*, also known as American hawthorn and May-apple, is found in eastern Canada, as are several ornamental hawthorns. (Contrary to its name, May-apple bears fall fruit.) The fruits are tasty, and it was while collecting them to eat that I decided to make a dyebath from some that were over-mature. I soaked out the fruits for two days, by which time the water had taken on a bright yellow-orange colour.

Parts used: mature fruit (September on). Grae (p 166) gives a recipe for using hawthorn blossoms, but her other 'hawthorn' recipes are for the species *Raphiolepis indica*.
Processing: crush fruits and soak them out in water to cover for two or three days. Stir this mush often. Bring it to a simmer, cook for half an hour, and then strain off the mush. Make certain you allow all the dye liquor to drip through the fruit mash (leave it to sit overnight).
Colours obtained: soft rich yellow (alum); strong medium gold (chrome); bright yellow-orange (alum, bloomed in tin); khaki (iron)
Fastness: excellent
How to identify: Hosie; Petrides; Saunders

Availability: from SH as nursery stock. Look for hawthorns along pastures and windrows. Its thorns are sharp and about 1 to 2 inches (2.54–5.08 cm) in length. Dyers may confuse it with mountain ash, which is superficially similar in that it too has red fall fruits. However, mountain ash fruits grow in clusters and those of the hawthorn are singular.

Heal-All

perennial weed, wildflower *Prunella vulgaris*

Heal-all is a short, purple-flowered weed with a squarish flowering head. It grows abundantly in hayfields and pastures. When our hay is cut, the heal-all grows up again within two weeks, giving the clipped fields a purple appearance.

Parts used: whole plant (flowers are too small to use alone unless you are dyeing a very small amount of fibre)
Processing: as for whole plants
Colours obtained: soft yellow (alum); medium brown (blue vitriol); orange-gold (chrome); medium grey (iron); brilliant gold (tin)
Fastness: excellent

How to identify: Cunningham; Frankton and Mulligan; Martin; Peterson
Availability: common across Canada and south into the United States
Special notes: Grae notes that the plant occurs in California, and she obtains a bright olive-green from it using an alum mordant (p 108).

Hemlock

coniferous tree *Tsuga canadensis*

The hemlock is a graceful tree whose branches droop slightly, making it appear quite delicate when compared to, say, a spruce. The tree does not usually grow perfectly straight, but has a somewhat 'leaning tip' (Saunders 29). Its dark green needles are stalked, which is a useful means of identification. Hemlock was first pointed out to me by Wilmot, NS, dollmaker Leila Banks on a hike we took through the fall woods. I then discovered, to my pleasure, that our land has several almost pure hemlock stands. Once you sort out fir, hemlock, and spruce, you will be able to identify each one. The needles of each are different, but as always, this is best seen using actual specimens.

Parts used: the fresh tips of hemlock boughs; bark
Processing: for tips, see fir (p 142). Treat bark as for any other bark.
Colours obtained: tips, collected in mid-summer: bright yellow (alum); strong brilliant yellow (tin); bright gold (chrome); warm brown (chrome and iron). No other yellow is quite as distinctive and clear in tone. It is a shade that is perfect for topping with indigo for green. The bark gives reddish-browns.
Fastness: excellent
How to identify: Hosie; Petrides; Saunders. Learning to identify the conifers in winter is perhaps easier for most people. The lack of competing foliage makes the outline of the coniferous trees more clear. Ask anyone you know who spends time in the woods to point out hemlock.
Availability: from SH as nursery stock, and some local nurseries. It can be transplanted from the wild, preferably in spring. Hemlock occurs in mixed woods, sometimes in small, pure stands (Annapolis, Hants, Halifax counties of Nova Scotia).
Special notes: Adrosko refers to hemlock being used to tan leather in Nova Scotia (p 42). Saunders mentions this as well, saying tanners would strip the bark off the trees and then leave them to rot in the forest (p 30). This practice probably contributed to their decline in numbers. Saunders also describes a red dye made from an extract of the middle bark of hemlock (p 30). Dyers wishing to try this should use bark from

felled trees only. Look for windfalls after a severe storm. I obtained a reddish-brown from the middle bark with alum and chrome.

Herbs

various species

Look under separate headings for chives, dill, parsley, and so on.

All herbs can be used to make dyes. Two useful publications are: *Growing Savory Herbs*, Publication 1158 (1963), Canada Department of Agriculture and *Herbs in Ontario* (see p 87).

Hollyhocks

garden flower, perennial *Althaea rosea*

Hollyhocks are tall-growing (5–8 ft, 1.5–2.4 m) perennials that were a stand-by in our grandmothers' gardens. They are currently enjoying a revival in popularity, and many dye references describe obtaining a red from the red flowers. Dawn MacNutt obtained a soft mauve from red blooms with an alum mordant. It may well be that the soil and climate in which hollyhocks grow contribute to one's success in obtaining the red. Lesch reports getting orange and rust with chrome from a bath made up of mixed colours of blooms (p 67).

Parts used: fresh flowers
Processing: as for flowers
Colours obtained: see above. From mixed shades (but mainly magenta, red and orchid): soft pinkish-beige (alum); taupe (iron); warm tan (chrome); light green (blue vitriol); blue (from red blooms, with alum and tin) (Dawn MacNutt's experiment)
Fastness: excellent
How to identify: hollyhock is pictured in most seed catalogues.
Availability: as seed from CK, DO, SH, and V. Available as seed or bedding plants from local nurseries. Some varieties are treated as annuals.

Horsetail

non-flowering, perennial *Equisetum arvense*

The common field horsetail is an unusual plant. It reproduces by spores, and when eaten by horses can be deadly. *Equisetum* actually produces two different stems: the hollow, spore-bearing stem appears first, as early as April. This stem is beige to brown and looks like a little pipe. Several weeks later the foliage stem appears. This has branchy bright green sprigs which give the plant the appearance of a prehistoric botanic form. Once identified, you will always know horsetail. It grows in poor soils, along roadsides, ditches, and driveways, and may even invade choice cultivated areas such as your flower garden. So prolific is this plant that several buckets full of the foliage can be collected in an hour.

Parts used: green foliage which appears following the spore stem
Processing: as for fresh leaves
Colours obtained: soft green (alum and blue vitriol); light green (alum); yellow-green (blue vitriol); bright dark yellow (chrome); brilliant medium yellow (tin); grey-green (iron). *Equisetum* is a fine dyeplant, giving a very wide range of strong, fast colours. Experiment!
How to identify: Cunningham; Frankton and Mulligan; Martin. This is one of the very few plants used for dyeing which is not described and illustrated in the Peterson reference.
Availability: common throughout, as far north as the Yukon (Cunningham 11). Look for horsetail in poor soil where it is quite dry (on the banks and sides of ditches, for example). Watch for the pale-coloured 'pipe' stems in April; within several weeks there will be an abundance of green foliage. It may be picked then, or left until it grows higher. Maximum height of the foliage stems is about 10 to 12 inches (25.4–31 cm).
Special notes: the shades obtained from horsetail make it one of the finest early season dyes to use. Take great care NOT to leave a basket of it around your back porch if you have grazing animals loose.

Houseplants

various species

Among my houseplants are several which require frequent pruning during their peak seasons of growth. These are the ivies (*Hedera*), wandering Jews (*Zebrina*), and philodendrons (*Philodendron*). German and Swedish ivies tend to grow prolifically, and since stem cuttings are so easily rooted there is always enough on hand to start a new plant or two. I found cuttings from a broken Swedish ivy (the hanging pot fell down!) yielded a brilliant orange when soaked in water for two days. However, I was unsuccessful in getting that orange on some fibre. Tin gave a bright yellow, and chrome a gold. Try using clippings from several different species in a single bath. Visit greenhouses and florists shops to collect damaged foliage plants or clippings.

Hydrangea

flowering shrub *Hydrangea*

Hydrangeas are flowering shrubs or small trees planted as ornamentals owing to the beauty of their enormous flower clusters. Popular as cemetery plants, hydrangeas come in shades of white, ivory, pink, and even blue. Those with blue flowers do, however, depend upon an acid soil to maintain their unusual colouring. If the flower heads remain on the shrubs through fall, they change in colour from, say, white, to dull bronze and then a rusty pink. Some turn from pink to bronze and then a tan. In any case, this is an outstandingly beautiful shrub and most gardeners would be horrified to hear that the fresh blooms give a good dye. Try collecting them from cemeteries (after seeing the caretaker). Most hydrangeas there are quite large, and have so many blooms a few would not be missed.

Parts used: fresh flowering heads
Processing: as for flowers
Colours obtained: from white and pink flowers; beige with alum; medium warm brown with chrome; grey with blue vitriol. Blue flower heads give yellow with alum and yellow-green with blue vitriol.
Fastness: excellent
How to identify: Petrides; Sherk and Buckley
Availability: from MC and SH as nursery stock. Available from most local nurseries which carry ornamental shrubbery.

Indigo

tropical herbaceous plant *Indigofera tinctoria*

Indigo is included in this book, even though it is not indigenous, because knowledge of this blue dyestuff is important to all dyers. There are many recipes for using indigo. Because the powdered dyestuff is insoluble in water, it requires special treatment. The availability of good, fast blues is important to the dyer, and indigo blues are just that. The powdered dyestuff is expensive, but a little goes a long way. The recipe offered here is by no means the only one a dyer should consider using. However, it has the advantage of being 'tried and true.' I have used this for workshops and other demonstrations where it is crucial to save time and yet explain the entire process with its many variables.

Two chemicals are required to make an indigo bath using the recipe given here. They are: sodium hydrosulphite (*Hydros*), and sodium hydroxide (common lye). Because of the similarity in their chemical names, the dyer is advised to refer to them as 'hydros' and 'lye.' Both chemicals are POISONOUS, very strong, and dangerous if inhaled or splashed on the skin. Always use rubber gloves when dyeing with indigo and take great care when cleaning up. Always add the chemical to the water. NEVER stir the water into the dry chemical as this produces toxic fumes.

Equipment: hydros and lye, powdered indigo
small enamel pan with a lid
2 large covered jars (pickle jars, etc.)
stirrers (several)
measuring cups, measuring spoons
thermometer
dyepot, fibre to be dyed

Processing: The following three mixtures are made up before the actual dyeing begins: Solution 1, Solution 2, and the indigo stock solution.

Solution 1: lye solution

To one pint of water in a quart jar (.5 l in a 1-litre jar), slowly add one-third cup (70 ml) of lye. Stir carefully. Cover the jar with the lid, label it, and set aside. (This lye solution is strongly alkaline and becomes quite hot, so wear gloves when preparing it or the jar will be too hot to hold.) Solution 1 will keep for three or four days, after which time it loses strength.

Solution 2: hydros solution

To one pint of water in a quart jar (.5 l in a 1-litre jar) slowly add 6 tablespoons (80 ml) of hydros. Stir carefully until the chemical is dissolved. This mixture is quite foul-smelling, so cover the jar tightly and then label it. Before you set it aside, you can wrap it in a plastic bag to diminish the odour. Hydros is a reduction agent. It reduces the indigo by removing the oxygen from the indigo bath.

Indigo stock solution
½ cup (120 ml) of Solution 1 (lye)
½ cup (120 ml) of Solution 2 (hydros)
3–4 Tbsp (45–60 ml) of powdered indigo*

*When ordering powdered indigo, specify that you do not want 'synthetic' indigo. Genuine powdered indigo is treated with chemicals to make it suitable to use with this recipe (see p 158).

You will need the indigo stock solution (as prepared above) plus more of Solution 2, hydros.

Method: heat 4 gallons (18 l) of water to 120°F or 48°C in the dyepot. Then add 4 Tbsp (60 ml) of hydros, Solution 2, to remove the oxygen present in the water. STIR THIS GENTLY, if at all, because you do not want to incorporate even more oxygen! It may be sufficient merely to tip the covered dyepot gently back and forth to distribute the hydros. Now let the water and hydros stand, covered, for 10 to 15 minutes to allow it to settle. Next add 4 to 8 Tbsp (60–120 ml) of the indigo stock solution. (Using a smaller amount will give light blues on 1 lb [453 g] of fibre, while the larger amount will give medium to dark blue with the same amount of fibre. Cover the bath, and again allow it to sit undisturbed for 20 to 30 minutes to allow the chemicals and indigo to settle. During this time the temperature should be maintained at 120°F or 48°C. You can wrap an old blanket around the dyepot. The temperature may fall a few degrees, but this is all right as long as it does not exceed 120°F or 48°C.

Dyeing: note: The bath will now appear to be a yellowish-green. This is normal.

Enter the wet fibre by easing it into the pot. Take care not to plunge it in and thereby introduce bubbles (oxygen). Once the fibre is in, replace the lid and put the pot on the heat. Always keep the temperature below 120°F or 48°C. (This is a conservative figure: some dyers consider a temperature of 140°F or 60°C safe, but I find novice dyers do well to stick to the lower temperature.) Allow the first skein or skeins to remain in the bath for 20 to 30 minutes. Take out the fibre and oxidize it.

Oxidization: this is what turns the fibre blue, or actually, more blue. Hold the dripping skein OVER A BUCKET (so as not to introduce additional oxygen to the bath) and allow it to drip for several minutes. Within 30 to 60 seconds it will turn blue, a colour much more definite than it was in the pot. Turn the skein one way and then another so all of it is oxidized. If the colour is satisfactory, hang the skein up to drip and label it (first dip, etc.). It may be redipped later on or, if you want a dark blue, returned to the pot immediately. Monsarrat, WI, student Doreen Lindo achieved stronger blues when she oxidized indigo fibres in sunlight, outdoors, for ten to twenty minutes between dips.

Control of colour: the colour given by an indigo bath is governed by the amount of stock solution used, the time the fibre is in the bath, and the number of dips. As a dyer gains experience, the colours can be predicted with a fair degree of accuracy, based on your previous records. Indigo blues are beautiful as they are, but extremely handsome when topped with strong yellows to produce greens. Experiment! No other dyestuff is quite so time-consuming and yet so rewarding.

Neutralizing: after each skein has been removed from its FINAL dip, it must be neutralized (because the indigo bath is strongly alkaline).

1 Rinse the indigo-dyed fibre in lukewarm water to which 1 cup (240 ml) of vinegar has been added for each pound (453 g) of fibre.
2 Rinse next in cool water.
3 Rinse a third time in a lukewarm bath using a non-detergent soap.
4 Rinse a fourth time in clear, cool water and hang to dry.

Drying: to avoid streaking indigo skeins, turn the fibre around often as it is drying. This prevents excess moisture from dripping onto one part of the skein. Drape the skeins over a plastic-covered rod or plastic clothesline. The dye may stain a painted wooden railing on a porch or fence.
Disposal of the indigo bath: because the lye and hydros in the indigo bath are strongly alkaline, it will not harm the plumbing to pour it down the sink. After all, many people use lye to unclog drains. But the indigo solution is messy; its purple slime clings to whatever it touches. Rural dyers with septic systems may wish to dispose of the indigo bath by digging a hole and pouring it into the ground in a spot where children and animals cannot get at it.
Cleaning equipment: all equipment used in indigo dyeing must be carefully cleaned, as even minute traces of blue residue will spoil a subsequent non-indigo bath. Throw the stirrers away, unless they are glass or plastic. Scrub all utensils with warm water and detergent soap. Then rinse them with a chlorine solution (p 15) followed by a clear water rinse.
Bottoming and topping with indigo: indigo-dyed fibres can be over-dyed with other dyestuffs to obtain good greens and already dyed yellows can be made green by subsequent dips in the indigo bath. 'Topping' usually refers to the process of covering a light shade (yellow) with a darker one (blue). 'Bottoming' is the reverse: the dark shade is dyed first and the lighter one goes on top. It is a matter of personal preference which method you use to obtain greens (or any other colours) from indigo. The equality of the resulting colour, however, depends on the clarity of the yellow dye. Good strong yellows for this purpose are: marigold, onion, apple bark, and hemlock tips. Goldenrod yellows tend to give bluish-greens when topped with indigo; the colours are unique and extremely attractive. Among the finest indigo dyes I have seen were those made by the late Edith Bethune, who lived in Berwick, NS. (Her daughter, Barbie Bethune, allowed me to look through her mother's dyeing records and swatch books.) Although the recipes are incomplete, her samples show an incredible diversity in colour, ranging from a clear turquoise to a variety of navy and dark blues. The turquoise was so bright and sparkling that I was amazed to learn that it had been dyed more than twenty years ago. Granted, the yarn sample books are kept in a dark place, covered with plastic. Even so, Miss Bethune told me her mother took

them with her as she toured the region giving public demonstrations, so these yarn samples were exposed to light some of the time. The turquoise carried only the brief description that it was made using a goldenrod yellow topped with indigo. Dyers who enjoy experimenting will do well to try indigo on brown and grey yarns. Try topping soft logwood rose-greys with indigo for purple heather effects.
Availability: see suppliers, page 228. Ask for genuine powdered indigo rather than the synthetic indigo, which is sufficiently different to require other recipes. The indigo plant is not useful for dyeing unless the dyer knows how it must be prepared before it can be used. This procedure is complex, so dyers are advised to buy the dyestuff in powdered form. Powdered indigo has added to it those chemicals necessary to render it soluble in water with the use of hydros as a reduction agent.
Other blues: see copper penny blue (p 127) and woad (p 220). Indigotin is the substance common to both indigo and woad. However, the indigo plant contains a larger amount of indigotin and hence renders faster, deeper blues.

Iris

domestic and wild flower *Iris*, various spp.

The yellow-flowering iris, whose roots yield a black, is the species referred to by Winnifred Shand as *Iris pseudacorus* (*Dye Plants and Dyeing* 63). The so-called blue flag iris grows wild in much of eastern Canada and the United States. It is *I. prismatica*. Among the domestic species are the bearded iris, *I. pogoniris*. Peterson describes *I. pseudacorus* as occurring throughout the region but only here and there, where it has escaped from gardens (p 100). Roland and Smith (p 216) report that this yellow iris can be found near Yarmouth and northeast Margaree in Cape Breton. I have used only the wild blue flag and its domestic counterpart. Most references make no mention of the root for dyeing, with the exception of Shand. The Krochmals (p 110) use the blossoms of an unspecified species for green and Grae gets blues and purples from the flowers of an unspecified blue-purple species (p 124). The rhizomes Shand refers to as the source of black are the underground rootstocks of the plant, which are white in colour!

Parts used: blossoms from any species; roots from yellow iris, *I. pseudacorus*, if available

Processing: treat blossoms as for fresh flowers. The roots, because they are tough, should be chopped and soaked out in water to cover before processing.
Colours obtained: read above. Blue flag rhizomes: yellow-tan (alum); soft grey (iron); blue flag blooms: yellow (alum); yellow-green (blue vitriol); grey (iron). The dyebath appeared to be blue but I was unable to affix this colour to the yarn with any of the usual mordants.
Fastness: good; rhizome colours as above. The grey I obtained from the root did not turn black with the addition of more iron in a very strong bath.
How to identify: Cunningham; Peterson. Wild food enthusiasts should be aware that iris can be mistaken for *Acorus calamus*, sweet flag, which is edible. But iris roots are poisonous (MacLeod and MacDonald 127). Most seed catalogues have photographs of irises.
Availability: from CK, DO, MC, and SH as bulbs (rootstock). Iris bloom in the Maritimes from early June on. Fields of wild blue flags can be seen as you drive along country roads. They prefer a damp, swampy habitat, and grow in wet spots in pastures, meadows, and in ditches.

Ivy

climbing vine *Hedera helix*

This is the common English ivy that is planted outdoors as a vine to cover porches, chimneys, and stone walls. (See houseplants, p 153, for indoor ivies.) The vine turns scarlet in the fall. Robertson gives a recipe for the berries, but does not mention the foliage (p 42). *Hedera helix* is not hardy in much of Canada, but Boston ivy (*Ampelopsis veitchi*) is. The *Hedera* I collected and used was from Pennsylvania, but I have not tried *A. veitchi*.

Parts used: fresh leaves; process accordingly
Colours obtained: soft yellow (alum); bright yellow (tin); tan (chrome)
Fastness: good
How to identify: Krobel; Petrides
Availability: as rootstock from MC and SH; available at most local nurseries
Special notes: Frankton and Mulligan, Hosie, Knobel, and Petrides illustrate poison ivy. Although it is quite another species, *Rhus radicans*, dyers who are learning plant identification would do well to know poison ivy.

Joe-Pye-Weed

perennial weed, wildflower *Eupatorium maculatum, Eupatorium perfoliatum*

Joe-pye-weed and boneset are tall plants with pinkish purple flowering heads. *E. maculatum* (Joe-pye-weed) has a flat-topped flower cluster, while *E. perfoliatum* (boneset) is actually a colour form (*purpureum*) of the usually white boneset. Neither plant is common in the Maritimes, but *E. maculatum* has a wider range, occurring in Newfoundland and west to British Columbia (Roland and Smith 655). There is some discrepancy between the information offered by Roland and Smith and Peterson (p 298), and dyers who document their work and thereby rely on proper identification are advised to read both references. The leaf shape and habitat serve as identifying characteristics which may help you decide which *Eupatorium* you have collected.

Parts used: fresh flower heads or whole plant
Processing: according to type (note: as some *Eupatorium* is localized and rare in certain areas, do not collect all the flowers from a clump. Leave some to reseed the patch. Many perennials weaken or winter-kill and are only re-established by self-seeding the previous fall.)
Colours obtained: from an unidentified species with purple flat-topped flower clusters, gathered in Hants County, NS: medium yellow (alum); bright gold (chrome); sharp yellow (tin). Whole plant: yellow-green (alum); green (blue vitriol and iron); grey (iron)
Fastness: excellent
How to identify: Martin; Peterson. Although Joe-pye-weed is not common, it is easy to remember once identified. The flower tops resemble a pinkish-purple goldenrod, which is similar in height.
Availability: Eupatorium prefers a moist soil, and grows along stream banks and the edges of swamps and bogs. I have picked *Eupatorium* in a meadow, adjacent to a small swamp. It blooms from late July to September.
Special notes: the unpicked flower heads turn soft grey in September and are collected for indoor flower arrangements. They take on a fuzzy appearance, which is a distinctive identifying characteristic. Davidson gives a recipe for *E. purpureum*, which may or may not be the variety I used (p 12).

Juneberry

shrub, small tree *Amelanchier*, various spp.

Juneberry is a small tree which grows in clumps. In the spring it has fragrant white blossoms which later turn into purplish-red fruit that are unequalled, to my palate. (Artist/decoy-maker John McClelland introduced me to juneberry in the form of a delicious jam he made.) Unfortunately for the beginning dyer, juneberry has a proliferation of common names. Here are a few: serviceberry, shadberry, wild pear, and bilberry. My father called *Amelanchier* 'Indian pear.' In any case, the tree is useful as a food source as well as for the various dyestuffs it provides. Hybridization among the many juneberry species is common (Roland and Smith 435) so exact identification may be almost impossible unless you take a specimen to a botanist. The shrub, or tree, is extremely common in most of Canada and the north-central United States. It grows along the edges of fields and pastures, and in windrows, amid alder, birch, mountain ash, and chokecherry.

Parts used: fresh leaves, fresh blossoms, berries (not tested)
Processing: process leaves accordingly. The blossoms were soaked out in water to cover and left for two days. I did not strain these off before entering the wet fibre in the bath, but used the 'contact' method of dyeing. That means that the dyestuff and fibre are processed together, a technique often used with lichens.
Colours obtained: from the blossoms: a warm brown (alum and tin); lemon yellow (alum); soft green (blue vitriol). From the leaves: yellow-green (alum); medium green (blue vitriol); brown (chrome); grey (iron); brilliant yellow-orange (tin)
Fastness: excellent
How to identify: Hosie; Knobel; MacLeod and MacDonald (they also use the common name 'Indian pear'); Petrides; Saunders; Sherk and Buckley
Availability: SH offers ornamental shrubs and tree forms of *Amelanchier*. Look for it growing in cut-over areas and along roadsides. The shrub is easy to transplant from the wild. The blossoms, although white, may take on a pinkish appearance as the immature leaves around them on each stalk are often purple-red in colour.

Knapweed

perennial weed, wildflower *Centaurea nigra*

Knapweed is also known as black knapweed and hardhead. It resembles a thistle, but has no thorns, and grows about two feet in height (.61 m) in hayfields, from mid-July through September. The flowers of knapweed are purple, but the base of the globular flower head appears black. Look for knapweed among red clover, daisy, buttercup, and vetch.

Parts used: flowering heads or whole plant
Processing: according to type
Colours obtained: (flowers): yellow (alum); yellow-green (blue vitriol); gold (chrome): (whole plant): greenish-grey (iron); tan (blue vitriol)
Fastness: excellent
How to identify: Frankton and Mulligan; Martin; Peterson
Availability: see above

Lamb's Quarters

weed, annual *Chenopodium album*

Lamb's quarters is a very common weed, one of the most abundant throughout the agricultural regions of the country (Frankton and Mulligan 48). It has flowers so small that they are inconspicuous. The entire plant – leaves, tiny flowers, and seeds – is a greyish-green. Lamb's quarters is an annual and grows as high as five or six feet (1.5–1.8 m). Wild food enthusiasts enjoy the leaves of lamb's quarters and MacLeod and MacDonald write that Indians used the seeds as grain (p 40).

Parts used: whole plant
Processing: as for whole plants
Colours obtained: yellow-tan (vinegar); soft yellow (alum); soft green (blue vitriol); grey (iron); tan (chrome)
Fastness: excellent
How to identify: Frankton and Mulligan; Gibbons; MacLeod and MacDonald; Martin; Peterson; Stewart and Kronoff
Availability: Look for lamb's quarters in cultivated soil (near gardens), waste places (ditches, roadsides), and near new house sites. Because it spreads by seed, where you find one plant you will usually find many more.

Larch

coniferous tree, deciduous *Larix laricina*

The larch is unique among conifers. It has needles like the fir, spruce, and pine, but loses this foliage just as hardwood trees lose their leaves. This unusual characteristic makes the tree easy to identify, especially in the fall when its needles are gold and bronze. The larch has two other common names: tamarack and hackmatack. The inner bark of this species is an unusual shade of purplish-red.

Parts used: fall needles (gold); fresh twigs; cones; bark
Processing: soak out needles and cones as you would the bark
Colours obtained: needles give a warm yellow tan with chrome; fresh twigs give a medium brown with chrome; the cones give tan with vinegar and salt; the inner bark gives a warm grey with an iron mordant. This grey had a distinctly purplish tone.
Fastness: good for all shades tested
How to identify: see above. Look for larch in wet areas such as bogs and marshes. Once the deciduous trees have lost their leaves, the gold and bronze needles of the larch are very conspicuous. They remain on the tree until November. Hosie; Petrides; Saunders
Availability: from SH as nursery stock. Larch are easily transplanted from the wild, but need a similar habitat.
Special notes: Davenport lists a most unusual shade from larch bark. She writes that it gives a lime-green on linen (p 117).

Lettuce

leafy vegetable *Lactuca*

Usually it is advisable to use fresh foodstuffs for dyeing only if they are unsuitable for human consumption. It was an interesting experience in consumerism when students of mine discovered that many supermarkets throw away an amount of lettuce each day that would probably feed many families. The outer leaves of lettuce heads are discarded if they are wilted or tinged with brown. They are also removed if the head is too large to fit inside the plastic display bag. While such wastage is deplorable, the students were able to obtain enough lettuce for a number of dyebaths, and the resulting colours were a surprise.

Parts used: leaves from any species of lettuce (may be wilted or otherwise unsuited for eating)
Processing: as for fresh leaves, but keep the temperature below a simmer to get greens
Colours obtained: soft yellow (alum); gold (chrome); green (blue vitriol); avocado-green (iron); bright yellow-green (tin)
Fastness: excellent for non-greens; the greens greyed a little with time

Availability: as seed, from all seed suppliers. Lettuce is easy for even a beginning gardener to grow. Visit local supermarkets for discarded lettuce leaves.

Lichens

non-flowering plants various genera and species

Lichens are non-flowering plant forms which many people wrongly refer to as 'moss.' (This description is basic. For more scientific information, dyers are advised to consult the Mason Hale book, *How to Know the Lichens* [see p 88].) Classified as cryptogams, lichens belong to a class of lower plants that includes algae, fungi, and the mosses. The lichens themselves are actually dual organisms composed of two plants, a fungus and an algae. These organisms live together in a relationship called 'symbiosis.' Slow-growing, long-living, and hardy in climates where other vegetation does not take hold, lichens are conspicuously absent from regions where the air is polluted. Because of this, lichens are an advertisement for clean air. Where they flourish, the environment is still relatively unspoiled. In addition to the information provided by Hale's book, dyers are advised to consult *Lichens for Vegetable Dyeing* by Eileen Bolton and an article by Marie Aiken of Gravenhurst, Ontario, called 'Lichens as a Dye Source.' (This article was originally presented as a paper to the World Craft Conference at Dublin, Ireland. It was reprinted in the spring 1971 edition of *Craftsman/L'Artisan*.) The article is important, as Ms Aiken describes in it her now-standard test to determine the presence of orchil in lichens (lichens containing orchil produce red).

Lichen types

1 Crustose, or crustaceous: crustaceous lichens look like disc-shaped crusts which form on rocks and trees. They usually have no lower cortex layer and appear to fade into the material upon which they are growing. Successful removal of a crustaceous lichen from the substratum for dyeing is almost impossible. Most lichens of this type are mineral grey, grey-blue, and orange-yellow in colour. Few crustaceous lichens are suitable for dyeing because of the difficulty in collecting them. This type is not extensively covered by the Hale book.
2 Foliose: Most of the common dye lichens are of this type. Foliose lichens are flat and often circular, with wavy or crinkled edges. The upper cortex layer is different from the lower surface of the thallus. Most foliose lichens are loosely attached to the surface upon which they grow. *Lobaria pulmonaria* is such a lichen. It grows on the trunks of a

variety of hardwood trees, and in wet weather may be quite easily removed from the bark using a spoon or knife. Others, like the umbilicates, are attached to the substratum by a single 'umbilical' cord. These, too, are easily collected, preferably in wet weather when they are soft. Foliose lichens are leafy in appearance, and lobed or branched. They grow on trees, rocks, and soil. (All the umbilicates listed in this book grow on rocks only.) Most are different colours wet than dry. Dry colours are grey, slate grey, charcoal, dull brown, and tan; wet colours are olive-green, khaki, and black.

3 Fruiticose: These lichens are shrubby and hair-like in appearance, or may take the form of small 'cups' or 'trumpets.' Some are erect, while others are pendulous and beard-like in growth. Many are too small to be worth collecting for dyeing, but others occur commonly and in abundance. Fruiticose lichens have a diverse habitat: they grow on trees, rocks, fence posts, rotted wood, soil, buildings, and atop other lichens. (Note: Lichens are not parasitic.) They occur in a wide range of colours, including yellow, green, pink, grey, and combinations, such as green with red cups.

Parts used: all the fleshy part of the lichen (the thallus) is used. Remove all other attached organic material (bits of bark, rock, needles, moss, etc.) in order not to contaminate what are often very subtle shades.
Processing: there are two basic methods. One is the boiling water method, used to obtain all non-red shades. The other is the fermentation method, used to obtain red from the umbilicate lichens and some *Parmelias*. (*Parmelia rudecta* tested by Alicia Marr, Sackville, Halifax Co., NS, gave reds.)
Colours obtained: from the boiling water method: a full range of shades from beige to tan; light to bright yellow and yellow-green; orange to rust; medium to dark brown; soft green to grey-green; khaki to brass and bronze. (Although lichens are substantive, the use of mordants extends the colour possibilities.) From the fermentation method: pink, rose, orchid, orange-red, barn red, magenta, violet, and purple
Fastness: all non-red shades are fast, but the reds are highly variable. There is considerable disagreement among dyers as to whether or not the reds fade. I have *Umbilicaria* reds which were dyed two years ago that have not changed in colour. (These are demonstration skeins which, when not in use, are stored in a trunk.) But other *Umbilicaria*-dyed fibres I have are faded. The species of lichen was the same as that used for the fast reds. Perhaps it is advisable, then, to reserve the use of lichen reds for articles not in everyday use.
How to identify: Bolton; Cunningham; Erskine (See p 88. Dyers are advised that some of the lichens included in this pamphlet have since

been reclassified.); Hale; Shuttleworth and Zim. Various technical references are included in the bibliography (eg, 'Lichens of Cape Breton Island,' *Annual Report of the National Museum of Canada*, 1952–3; see p 234).

Availability: lichens are common all over this continent and, indeed, that is the problem! While some species abound, others are extremely rare, so correct identification is important. Reliance on common names is unwise. There are several foliose lichens that resemble 'lungs,' and many fruiticose types that look like 'beards.'

Habitat and locale of umbilicate lichens: umbilicate lichens are usually found growing on rocks, and are often referred to as 'rock tripe.' They cling to granite rocks in open woods or near the sea. Umbilicarias will grow on the top or horizontal surfaces of rocks but these lichens tend to be smaller than those found on the vertical surfaces. They look like small, leathery plates or discs, and in dry weather are brittle and crisp. In wet weather, the umbilicarias not only change colour but become soft, pliable, and rubbery. They should be collected when in this state. Often several species will intermingle on a single rock. The under surface (lower cortex) of some umbilicates are buff or fawn in colour and the upper has raised 'pustules' which are warty in appearance. *U. papulosa* is such a species. Others have a black lower cortex which looks velvety, like *U. pennsylvanica*. Look for umbilicate lichens in Nova Scotia in Shelburne, Queens, Halifax, and Guysborough counties. Also *U. muhlenbergii* was found growing in Victoria County, Cape Breton. (Identified by K. Sonnenburg, naturalist, Cape Breton Highlands Park. Further evidence of the presence of umbilicate lichens on Cape Breton Island is given in an article by I. MacKenzie Lamb, 'Lichens of Cape Breton Island' in *The Annual Report of the National Museum of Canada*, 1952–3, Bulletin 132. The species collected were *U. duesta* and *U. muhlenbergii*, in Inverness and Victoria counties.)

Habitat and locale of boiling water lichens: Lobaria pulmonaria, Usnea comosa, Peltigera apthosa, and *Cladonia alpestris* are among the most commonly collected lichens processed by the boiling water method. The common name will be given for each of these lichens only as a means to help those new to lichen identification. As soon as you have located the species, immediately learn and use the Latin nomenclature. No matter where you are in the world, a lobaria is always a lobaria.

Lobaria pulmonaria – Commonly called 'lung lichen' and 'oak rag,' this *Lobaria* species grows quite large and hangs, rag-like, from the trunks of maple, oak, beech, and less often, coniferous trees. (For a long time I doubted that *L. pulmonaria* grew on softwoods. I finally found some, in my own woods. The usual habitat is on hardwoods.) It is buff, tan, or khaki when dry, and ochre to olive-green when wet. The lower surface is mottled beige and white, and the entire thallus of the lichen has bumps which give it the appearance of lung tissue, but *L. pulmonaria* is so pleasant to say that there's little reason not to learn it. In eastern Canada and northern New England it occurs commonly, especially in mature stands of open woods. Look for *L. pulmonaria* in maple sugar country. In midsummer, it may be difficult to see, but it stands out on a wet day, appearing as bright olive-green patches well up on the tree trunks. You may need a rake to pull it down. Never take more than you need. A plastic bucket full will dye 4 oz (114 g) of fibre medium brown, and 8 oz (228 g) a warm golden-tan.

Usnea comosa – 'Old man's beard' is usually known even to non-dyers because it is extremely common. The usneas are not all the same colour, but *U. comosa* is usually light green, greyish-green, or greenish-yellow. It is branched and tufted, and may be erect on a tree branch rather than hang down like hair. If the usnea you collect is much longer in length and hairlike, it may well be *U. dasypoga*. Another characteristic which distinguishes *U. comosa* from *U. dasypoga* is that the former becomes black at the very base where the lichen is attached to the tree. Most usneas grow on coniferous trees. Sometimes a dead spruce will appear to be almost covered with usnea, giving it an other-worldly appearance and contributing to the myth that lichens are parasitic. The usneas are best if used in a strong dyebath. Robertson writes that *U. lirta* gives a purple with the fermentation method (p 108). I found no such species listed by Hale, but he does give a *U. 'hirta'* (p 186), which resembles *U. comosa*. I suspect that the 'lirta' mentioned in Robertson's book is a typographical error.

Peltigera apthosa – This lichen is listed by Cunningham (p 1) as one which occurs commonly throughout Canada, although I looked for a year before finding it. Once identified, however, it is unlikely to be confused with any other species. Commonly called 'dog lichen,' it grows to the size of a hand. The outer edges of the thallus turn up and over, revealing the buff colour of the lower cortex. When dry, *P. apthosa* is dark grey or a dull grey-green. Because it often grows on mossy beds under evergreen trees, it is almost impossible to see. But on a wet day it turns a shiny bright green and fairly shouts its presence. Superficially, *Peltigera* resembles *Umbilicaria* in that it is more or less flat and plate-like. But *Peltigera*'s curled-up edges are distinctive, as are the tiny cephalodia or warts on the upper surface. These are the size of large pin pricks. I have always located *P. apthosa* growing on mossy ground near conifers in mature, mixed woods.

Cladonia alpestris – Even though it is not a moss, this lichen is called 'reindeer moss.' The appearance of *C. alpestris* is unusual and distinctive. It forms stiff, upright tufts that look like miniature shrubs. This lichen is usually greyish-white or greenish-white in colour and grows on soil in

woods and rocky areas. In dry weather, it is stiff and crumbly, and when wet, soft and sponge-like. Usually *C. alpestris* forms large mats which cover several square yards of ground. It can be found most often near rock outcroppings and on the edges of mature, coniferous woods.
Special notes: it is significant to note that Davenport's book, *Your Yarn Dyeing*, published in Britain in 1955, was written before dyers Eileen Bolton and Seonaid Robertson established the fact that orchil-yielding lichens do, in fact, exist in the United Kingdom. In the chapter entitled 'Native Plants which Yield Dye,' Davenport states that reds and purples are unavailable from native lichens (p 123). Either she was referring to the reds and purples obtained by roccella, a Mediterranean lichen, or else she was unaware that the umbilicates and parmelias that occur in Britain contained orchil. This is puzzling, as her research was obviously extensive. But she ends her chapter by writing that she hopes further lichen investigation will occur; indeed, Bolton and Robertson carried on and today we all benefit from their research. (The late Violetta Thurston, another British dye expert, wrote her book around the same time, or possibly before, Davenport. Thurston acknowledges the orchil lichens and gives recipes.)

GENERAL PROCESSING FOR ALL LICHENS

To release in the dyebath all the pigment contained in the inner layers of the thallus, lichens must be thoroughly torn, crushed, or shredded before using. If you have collected the lichens when wet, they will be too rubbery to rip apart. Let them dry first by spreading them out in the sun. Dry lichens weigh very little and may blow away, so cover them with a piece of netting or a screen. Lichens can be dried inside by laying them on newspapers in a warm spot and leaving them for several days or until they are crisp. Once dry, crush them with your hands, as finely as you can. Dry lichens in this state can be kept several years. Store them in an uncovered container, a burlap bag, or old pillow case. While dry *Umbilicaria* may be kept indefinitely, most foliose and fruiticose lichens are thought to yield the best colours if used when fresh.

BOILING WATER FOR LOBARIA PULMONARIA, USNEA COMOSA, PELTIGERA APTHOSA, AND CLADONIA ALPESTRIS

Unless otherwise stated, a medium strength dyebath is used for boiling water lichens (equal weights of lichen and fibre).

Lichens processed in boiling water may be treated one of two ways. The lichen is soaked out in water to cover for two or three days, and then cooked out to produce the dyebath, or the lichen and fibre are layered in the dyepot together. This is called the 'contact' method and is a traditional Scottish technique (Robertson 103; Lillias Mitchell writes that this method is still used in Donegal, Ireland, and it is also used by some Cape Breton dyers). It is reputed to yield full, strong colours, but has the disadvantage of limiting the amount of fibre that can be dyed at one time, since the lichen takes up at least half the space in the pot. Also, small particles of the cooked lichen adhere to the dyed fibre even after it is dried. (Dyeing unspun fleece by the contact method can mean that the spinner later runs into these particles as the yarn is being spun.) Supposedly, a good shaking will loosen these, but my personal experience is that they stay attached. After dyeing, I wound the skeins into balls, wound a warp, and wove a fabric. The cloth was then washed, dried, and pressed. Small particles of the lichen could still be seen. My preference, therefore, is for the alternate method.

Soak the shredded, crushed lichen in water to cover for two or three days. Lichens swell up after they become wet, so add more water if it seems necessary. Stir the mixture several times a day to make certain all the lichen particles are thoroughly wet. The water will take on a strong colour within several days. Then proceed with the dyeing.

Cook out the lichens in simmering water for two or more hours, depending upon the desired colour. The shade can be checked periodically by dipping a pyrex cup into the dyepot and holding this to the light to determine what colour the bath will be. When the colour is satisfactory, remove the pot from the heat and strain off the dye liquor. Discard the cooked lichen after all the liquor has been squeezed from it. Then pour the dyebath back into the dyepot and add the wet fibre. If you have strained the liquor off carefully, there should be almost no lichen particles left in the bath.

Many dyers leave the fibre to sit in the lichen bath overnight before putting it on the heat. Others process the dyebath, and then leave the fibre in the bath overnight. Both methods are thought to yield stronger colours. But lichen-dyeing techniques are as variable as lichens themselves. Improvise and experiment and then do it your own way. Beginning dyers will find that they soon come to love the characteristic smell of lichens cooking out in a pot on the stove.

UMBILICARIA FERMENTATION

This method is appropriate for those species of umbilicaria known to occur along the Atlantic coast, in the north-central United States, and along the shores of lakes Superior and Michigan, including *U. deusta*, *U. mammulata*, *U. papulosa*, and *U. muhlenbergii*. (Non-umbilicate lichens can be fermented in ammonia to yield reds. Some of these include *Parme-*

lia rudecta and *Evernia prunastri*. See Bolton 58.) During fermentation, the crushed *Umbilicaria* is steeped in an alkaline solution of water and ammonia. The traditional fermentation agent used was stale urine, and even today the use of urine in the bath may smell no more disagreeable than ammonia. The results obtained are virtually identical, although the ammonia shades are perhaps a little redder. The orchil contained in umbilicate lichens will produce red if fermented in an alkaline solution which is aerated daily to incorporate oxygen. Whereas the indigo dyer assiduously tries to keep oxygen out of the dyebath, the *Umbilicaria* dyer must attempt to get as much of it *in* the bath as possible. Without sufficient oxygen the orchil fermentation will not work properly, and dull brown rather than red will be the disappointing result. Most beginning dyers make the same mistakes when fermenting orchil lichens:
– the lichen is not sufficiently ground up;
– too little ammonia is used;
– the fermentation period does not last long enough;
– the fermenting mixture is not stirred often enough.

UMBILICARIA RECIPE

Note that this recipe is sufficient to dye 3–4 POUNDS (1.4–1.8 k) of wool fibre, either fleece or spun yarn. To dye smaller amounts of fibre, decrease the recipe accordingly.

4 cups (1.1 l) of *Umbilicaria*, dry (This measurement is taken after the lichen has been crushed up with the hands. If a blender is used to grind it finer, measure this amount BEFORE using the blender.)
2 cups (.5 l) of liquid household ammonia OR 1 qt (1.1 l) of stale urine
2–4 cups (.5–1.1 l) water (hard water often gives the best colours)

After crushing the dry lichen with your hands, it may be further pulverized by grinding in a blender or using a mortar and pestle. If the lichen is ground very finely, use the smaller amount of water given in the recipe. Put the lichen, water, and ammonia (or urine) in a large plastic container with a tight-fitting lid. A glass jug will do, provided the mouth is large enough to enable you to stir vigorously. Stir the mixture well for five minutes or until you can see that all the lichen particles are thoroughly wet. Cover the solution with the lid, making certain that it is on snugly. This is important, because you do not want the oxygen you have just stirred into the mixture to escape. Place the fermentation container in a warm, dark place where it is easily accessible and yet out of the reach of chidren and pets. (Marie Aiken advises keeping the container in a dark location, but Bolton places hers in the light. I prefer a dark location, as the reds produced seem stronger.) At this point, the lichen-water-ammonia mixture is dark brown, and it may remain so for a week or more. Eventually you will notice a slight reddish tinge in the solution, which gradually increases until the mixture is much like the colour of blood. Two things are important to remember: one, the bath must be stirred AT LEAST twice a day (more often if possible); and second, the mixture must not fall below 60°F (15°C) or get warmer than 90°F (33°C). There is no need to use a thermometer as long as you keep the bath at room temperature. However, if fermentation is carried out in the winter in homes without central heating, make certain the fermentation container is kept near a source of heat at all times. Even though the mixture smells, don't relegate it to the back porch or basement in the winter. If the solution evaporates over a period of time, add a little extra water and ammonia. If the red does not appear within one week (and you have been stirring VIGOROUSLY, several times a day), add more ammonia. Dyers who use *Umbilicaria* which has been fermenting for less than four weeks don't give it a chance to work and demonstrate its true colour potential. Yes, you can get pink and rose from a bath that is four weeks old. But from a bath that is eight or ten weeks old, the results are fine, strong reds and magentas. *Umbilicaria* dyeing is worth all the trouble it takes, but don't expect dramatic results when you have not had enough patience to let the bath mature.

To dye with the fermented lichen, you simply strain off the liquid from the plastic container and add sufficient water to it in the dye vessel to make a bath. The lichen itself SHOULD NOT BE THROWN OUT. You can add more water and ammonia and ferment it all over again, although this second batch may produce lighter colours. Remember that the orchil liquor is very strong, and the amount given in this recipe will dye many skeins of yarn. You may not wish to pour it all off at once. In that case, take a cupful of the mixture, recover the container, and let it continue to ferment. Add sufficient water per cupful of liquor to make a dyebath and proceed with the dyeing.

The reds obtained from orchil lichens depend on a variety of factors: the amount of lichen used; the amount of ammonia used; the amount of oxygen incorporated in the solution; and the fermentation time. If you maximize all these variables, you obtain the strongest, deepest shades. Although they are substantive dyestuffs, the use of mordants greatly increases the range of colours obtained from all lichens. Pre-mordanted fibre may be used, or you can add mordants to the dyebath as you proceed. The following colours are only approximations. There are many variables. No bath was aged for less than ten weeks. Alum and cream of tartar: turkey to barn red, first dips; medium rose to soft rose, successive dips; chrome: cardinal red, first dip; faded barn red, successive dips; tin: bright magenta, first dip; magenta-rose, successive dips; no mordant: orchid-rose, first dips; soft orchids, successive dips; vinegar: barn red, first dip; rose, successive dips; iron produces dull rose drabs, and reds bloomed with tin take on a sharp brilliance. True clear reds are difficult to obtain. In my experience, the best reds of this type came from baths processed in stainless steel dyepots using vinegar as a mordant. Experiment and record your results in order to repeat the process that, perhaps unexpectedly, gave you a superb red. Halifax, NS, dyer Sheila Simpson has obtained excellent results dyeing with umbilicate lichens using a method based on the solar approach (see p 234). She prepares the fermentation mixture as outlined on page 170. The fibre to be dyed is placed right in the fermentation container at that time, and after several weeks, using no heat or mordants, it takes on strong colours that seem more fade-resistant than heat-processed fibres. An alternate method she has used is to let the fermentation mixture sit (stirring only occasionally) for several months. The fibre is then put in the container and dyes very well within two days. These colours, too, were fast to washing and sun exposure (see p 196). Based on Sheila's work, I dyed ten pounds of yarn with *Umbilicaria* using no heat or mordants. Nor did I use the sun. Fibre was placed in the dye liquor for two to three days, after fermentation was over and the lichens were strained off. The colours were very fast and exceptionally bright.

TEST FOR THE PRESENCE OF ORCHIL

Orchil is that substance in the white fungal layer of the lichen thallus which produces red when fermented with water and ammonia. The test to determine if orchil is present in a lichen is credited to Marie Aiken (see p 164). With a sharp knife, scrape the upper cortex layer of a lichen until the white fungal layer is revealed. Apply a drop of common liquid household bleach to this white spot and if it turns bright red immediately, orchil is present. I have experienced difficulty when testing some umbilicates for the presence of orchil. A few did not turn red, but upon fermentation still yielded reds as anticipated. Try the test on more than one specimen of a species.

COLLECTING LICHENS

Although some dyers think it is important to collect lichens at specific times of the year, others collect them as required. I have used lichens harvested year round with little variation in the results. The conscientious dyer disturbs as little of the natural environment as possible when collecting lichens, at any time of the year. Such a person goes out to collect *Lobaria pulmonaria* and comes back with just that. If you are curious about other species, take specimens of each, to identify later. (Even if you take along your reference books, correct identification in the field may be impossible.) Collect on a wet or foggy day. Dry lichens crumble easily, and you may unintentionally dislodge more than you actually need. Collect from the most inaccessible areas you can reach. Gathering lichens in very public places creates curiosity about what you are doing. (Peggy's Cove, NS, and other tourist areas should be avoided.) Once you tell onlookers, they too may start to pick bagfuls of lichen that they have neither the skill nor the interest to use. Wear sturdy shoes (rocks are slippery when wet) and sensible clothing.

Dyers have an obligation to justify their work within reasonable limits. This is not difficult if you collect all dyestuffs with care and consideration. Taking any plant (even a specimen) from a protected area or a national park is against the law.

Lily

perennial, domestic and wild flower *Hemerocallis fulva*

The so-called day lily is the only tall-growing domestic and wild perennial that has a bright orange bloom of the characteristic lily shape. Day lilies were originally planted as garden flowers, but over the years they have escaped so that now they bloom in the wild along many rural roadsides. The flowers first appear in June, with each bloom lasting only a day. However, a large clump of *Hemerocallis fulva* will provide blooms for a continuous period of almost a month. The flowers and immature flowering buds are a favourite wild food.

Parts used: blooms, roots
Processing: according to type
Colours obtained: fresh blooms: yellow (alum); rust (alum and chrome); burnt orange (chrome and tin): roots: yellow-orange with alum and tin
Fastness: good
How to identify: CK, MC, and SH seed catalogues. Gibbons; Peterson; Stewart and Kronoff
Availability: day lilies are common throughout most of the eastern provinces and states. Look for large clumps near abandoned homesteads and turn-of-the-century gardens in towns and cities.

Special notes: Robertson reports getting a blue-green from fresh day lily blooms with blue vitriol and iron as mordants (p 45).

Linden

See basswood, page 98.

Lupin

perennial, domestic and wild flower *Lupinus*

It has always surprised me that so few dyeing references mention lupin as a dye source. The Threshes give a recipe (p 15), explaining that their samples did, however, change in colour over time. They call the plant 'lupine,' which is another spelling. Grae mentions obtaining blue and green from the flowering spikes but says the shade has only fair light fastness (p 94). I have used domestic Russell hybrid lupins and wild plants for dyeing for a number of years, and find my samples still have an excellent colour although the shade is now slightly different, more grey-green than green-yellow. I used lupin-dyed wool in a cape woven for one of my daughters, and a remnant of that fabric was made into a cushion that has been on my sofa for years.

Parts used: flowering heads, flowers and leaves, or whole lupin plant
Processing: according to type
Colours obtained: light-coloured blooms (white, pink, orchid) give yellow-green shades if just the flowers are used, and greens if the leaves are used as well. Blue and purple blooms give good green and grey-green with iron and blue vitriol. The whole plant gives yellow-green (alum) or green (blue vitriol, iron).
Fastness: see above. I think lupin shades are fast, but highly variable and quite likely to change in colour.
How to identify: CK, DO, MC, SH, and V seed catalogues: Peterson; ask locally. Farmers are especially anxious to be rid of this plant, as it sours cows' milk.
Availability: from above sources as seed or plant stock. Look for fields of lupins blooming from mid-June on. In rural areas, lupins often fill ditches with purple, blue, rose, pink, and white blossoms. The wild root-stock is easy to transplant if the deep roots are dug up when the soil is wet. Like day lilies, lupins seem to prefer to grow on the poor soil at the edges of country roads. Lupins were popular perennials in turn-of-the-

century gardens and clumps can often be found adjacent to old homesteads and deserted farms.

Maple

ornamental hardwood tree *Acer*, various spp.

There are many maples, most of them easily recognizable in the fall because of their distinctive bright red and orange coloration. A few of the most common species are: *A. rubrum* (red maple); *A. saccharum* (sugar maple); *A. platanoides* (Norway maple), and *A. saccharinum* (silver maple). Although I often use the leaves and bark of various species interchangeably, many dyers use a certain variety for specific results. Robertson uses the bark of Norway maple to obtain a pinkish-tan with chrome (p 54). The Krochmals use an unspecified species of maple bark to dye cotton grey with chrome and vinegar as mordants (p 251). Adrosko mentions the bark from *A. rubrum* as being used for a blue dye for wool and linen in the eighteenth century (p 43). Davidson writes that her grandmother got purple dye from rotted maple wood with iron added as the mordant (p 14). Saunders describes a recipe for making black ink from the bark of *A. rubrum* (p 78). This is also mentioned by Sterns in her article, 'Family Dyeing in Colonial New England' (in *Dye Plants and Dyeing* 80).

Parts used: fresh leaves; autumn leaves; bark (from firewood, deadfalls, or twigs)
Processing: according to type. For particulars regarding the colours mentioned above, consult the reference given.
Colours obtained: fresh leaves give yellow with alum; beige with vinegar; warm tan with chrome; grey-green with iron and soft green with blue vitriol. Fall leaves from *A. rubrum* give a warm, rich tan with chrome and tin in a strong bath. Bark from *A. rubrum* gave medium brown with chrome; bright dark brown with chrome and tin; rose-tan with alum; medium grey-brown with iron. Twigs of mountain maple (*A. spicatum*) gave a reddish-tan with alum and chrome.
Fastness: excellent for all
How to identify: Hosie; Knobel; Peterson; Saunders
Availability: from MC and SH as nursery stock; available from most local nurseries
Special notes: the blues and purples from maple bark warrant further investigation as they are mentioned in a number of references. *Foxfire 2* (p 211) gives a recipe which calls for using the inner and outer bark and boiling the dye for a day.

Marigold

annual, garden flower *Tagetes*, various spp.

The marigold and the petunia are, along with the geranium, our most commonly planted garden annuals. Distinguishable by its brilliant blooms and unattractive scent, the marigold is available in numerous sizes ranging in height from 5 inches (12.7 cm) to 3 feet (.92 m). The flowers are magnificent, and new two-toned hybrids complement the already extensive range of colours, which includes lemon yellow, bright yellow, yellow-gold, yellow-orange, bright orange, gold, ochre, rust, maroon, rust-maroon, and deep rust. Although marigolds reach the peak of their bloom late in the summer (August and September), they remain in flower in some locales well on into October. Gardeners who cover their marigolds on nights when there is a frost can prolong the bloom to early November. Once frost-bitten, the flowers yield wonderfully rich shades that defy mere descriptive names: rich ochre, khaki-olive, olive-green, avocado-green, and deep burnt orange. No other common dye flower is as adaptable to the dyer's needs. Only onion skins compare with marigolds in ease of preparation and availability. Beginning gardeners can grow marigolds from seed or bedding stock even in poor soil. The marigold will tolerate draught, partial shade, and full sun. The more you pick the blooms as they fade, the more marigold flowers. The popularity of marigold in planters around public buildings means that there is always a ready supply. Many dyers freeze the blooms to use them later in the season. (Eveline MacLeod, South Haven, Cape Breton, told me she froze the blooms but that the colours obtained were slightly darker than they would have been normally.) A motel at Bedford, NS, has hundreds of feet of garden planted with marigolds. Each autumn I have stopped there to ask permission to pick faded blooms, and each time I collect enough in five minutes to dye several pounds of wool yarn. Most gardeners are delighted to share their flowers with you, especially after the frost, but always establish a good rapport first by taking the advice they offer.

Parts used: the flowers, separated by colour or used all together, either fresh or faded; severely frost-bitten blooms are extremely useful to obtain special shades
Processing: as for flowers. Marigolds are not often used to obtain a clear, light yellow because their pigment is so strong; a bright but dark yellow more often results. A light yellow is more easily obtained from onion skins, although the use of all yellow fresh marigolds in a weak bath will give a lemon-yellow if the heat is kept down. Process fresh blooms until

the desired colour has been obtained. The timing is highly variable, owing to the large amount of pigment in the marigold. Keep checking the pot as the dyeing progresses to determine the colour. Frost-bitten blooms need not be torn apart. The longer the processing, the deeper and darker the colours will be. Do try dyeing with marigolds in pots of different metals, especially brass, copper, and aluminum, as the results are interesting and unusual.

Colours obtained: fresh blooms, light-coloured flowers (yellows): clear light yellow (alum, short processing, low heat); medium bright yellow (alum, longer processing); rich gold (alum, processed at high heat, with vinegar added to the bath). Fresh blooms, orange and rust flowers: orange (tin); burnt orange to rust (chrome, bloomed in tin); brown (chrome, saddened in iron). Frost-bitten blooms, mixed colours, in a copper boiler lined with tin: green (blue vitriol); brown (chrome); grey (iron). Faded blooms, dark colours, in a brass pot: olive-green (blue vitriol); khaki (chrome); bronze (tin)

Fastness: excellent for all shades, although the dark colours may change slightly over a period of time (an olive-green became brown-olive after a year)

How to identify: Herwig; ask locally

Availability: as seed from CK, DO, MC, ST, and V. Available as seed and nursery stock from nurseries and department stores. Marigolds planted directly outdoors from seed may not attain full growth. Most experienced gardeners start them indoors in March in 'jiffy pots.' Beginning gardeners should plant nursery bedding plants. These can be set out in May. Once marigolds are flowering, remove all faded blooms as they occur to encourage more to grow.

Special notes: The wildflower known as marsh marigold, or cowslip, is *Caltha palustris*. According to Roland and Smith (p 395), its habitat in Nova Scotia is very limited. Because it is rare, dyers wishing to try this plant are advised to order it as nursery stock from specialty houses such as CK. However, my attempts at planting such stock have been unsuccessful, and the plants are very expensive.

Mint

perennial herb *Mentha piperita, Mentha spicata*

Mint is a perennial herb which grows in the wild and is also cultivated for domestic use. Although rarely listed as a dye source, mint is an excellent dyestuff and easy to grow. There are many types, including *M. piperita* (peppermint) and *M. spicata* (spearmint). Peppermint has dark green leaves and reddish stems, while spearmint has lighter green leaves and tiny lavender flowers. You can usually smell wild mint in the early summer. It grows in fields and meadows – just follow your nose. Domestic *M. piperita,* once established in the garden, will rapidly take over, crowding out other plants. I locate mine away from the vegetables and flowers, at the edge of the hayfield, where it can grow freely.

Parts used: whole plant
Processing: as for whole plants. *M. piperita* and *M. spicata* can be used interchangeably, although the latter species does not seem to have as much pigment.
Colours obtained: soft yellow (alum); beige-tan (blue vitriol); gold (chrome); yellow-orange (tin); tan-grey (iron)
Fastness: good
How to identify: Erichsen-Brown; MacLeod and MacDonald; Peterson; Stewart and Kronoff
Availability: as seed, from CK, DO, MC, ST, and V. Look for wild mint in fields and meadows in early summer, growing amid wildflowers that prefer rich, damp soil. Where I live, spearmint grows along the roadside and in ditches. Once you plant *M. peperita* for domestic use, you will always have a supply. I cover mine with a mulch in late autumn, and uncover the patch in late April. Within several weeks there is enough mint for everyday use.

Mountain Ash

deciduous ornamental tree *Sorbus americana* and others

The two native species of mountain ash are *Sorbus decora* and *Sorbus americana*. *S. aucuparia* is an introduced species, European mountain ash. It is also called rowanberry and frequently planted as an ornamental because of the showy orange-red fruit clusters which develop in later summer. These fruits are in the form of large clusters and are a unique identifying feature. The mountain ash is usually not a large tree and may grow in a clump like the alder and chokecherry.

Parts used: fresh leaves; fresh berries or fruits (Although I find the berries give a mediocre dye, the leaves are excellent as a dyestuff.)
Processing: leaves as for leaves; fruit as for berries. Shades obtained from the mountain ash fruit are fast, but it is messy to use.
Colours obtained: golden-tan (alum); warm tan (blue vitriol); bright dark gold (chrome); soft grey-brown (iron); bright yellow-gold (tin). The berries give a salmon-pink with alum, and a rose-tan with chrome.

Fastness: excellent for dyes from leaves and fruits
How to identify: Hosie; Knobel; Petrides; Saunders
Availability: as nursery stock, from MC and SH and some local nurseries. Mountain ash is easily transplanted from the wild in spring. Look for it growing in windrows along the edges of pastures and hayfields. It is not easily confused with hawthorn, which also has red fruits, because the mountain ash fruits are in clusters and each berry is small. The hawthorn fruits occur singly and are the size of a small crab-apple.
Special notes: Although I have not tested mountain ash bark, both Davenport and Worst (p 40) give recipes, the latter stating the resulting colour as grey-red, which probably means a warm, rosy taupe.

Mullein

biennial, weed, wildflower *Verbascum thapsus*

Mullein is also known as the flannel plant because of the interesting texture and appearance of its velvety leaves. It usually occurs as a single specimen, growing in waste places and on poor soil to a height of 4 or 5 feet (1.22–1.5m). It has a single, club-like yellow flowering head and greenish-white leaves that clasp the stem near the bloom, but lie flat at the plant's base, like a flannel apron. No other plant looks quite like mullein, and once identified you will never mistake it for anything else. I find it attractive because of the distinctive habit of growth, and am loath to use a mullein for dyeing unless I can see other specimens growing in the vicinity. The root is extremely deep, but if you dig it up mullein can be transplanted to a convenient spot near where you live. Because of its enormous size, you do not need more than a few flowering heads or several handfuls of leaves to make a dyebath. Remember, as with all biennials, that the first year's growth produces no flower but only the characteristic flannel rosette of leaves. The stalk is so tough that you may need a small hatchet to chop it up for the dyebath.

Parts used: flowering heads; flowering heads and fresh leaves; flowers, leaves, and stalk. A single plant will dye 4 oz (114 g) of fibre.
Processing: because of its flannel-like texture, mullein must be thoroughly soaked out in water to cover before making the dyebath. The leaves take several hours to become wet, so it is important to stir the plant down several times a day, and let it soak out several days if time permits.
Colours obtained: soft yellow (alum); bronze (blue vitriol); gold-bronze (chrome); grey (iron); bright yellow (tin). The grey with an iron mordant is unusually beautiful. It had a slight yellowish cast that gave the colour warmth.

Fastness: excellent
How to identify: Martin; Peterson
Availability: see above. No other plant with a yellow-flowering head grows as tall on a single, unbranched stalk that is very thick. Occasionally you will find a group of mullein but often the plant grows singly.
Special notes: Lesch says that mullein harvested early in the season does not give good colours (p 47), but I find the mature plant too tough to process with ease, so often use plants gathered in July.

Mum

See Shasta daisy (chrysanthemum), page 206.

Mushroom

non-flowering plant various genera and species

Read about fungi, page 144.

For the purpose of this book, those fleshy fungi with gills are considered 'mushrooms,' as are such forms as the *Calvatia* (puffballs) and the *Morchella* (morels), although both of these are botanically distinct from gill-bearing fungi.

IMPORTANT: Take no chances when collecting mushrooms for dyeing! Do not take small children along. Have a friend accompany you. Collecting poisonous mushrooms is dangerous if you handle them with bare hands, so always wear gloves. Novice mushroom collectors should NEVER eat specimens intended for dyeing, especially if poisonous and non-poisonous varieties are carried in the same container. The best policy is to take a mushroom-collecting course or have an amateur mycologist accompany you. (Courses are offered, usually in the fall, by YMCA's, YWCA's, universities, and naturalist and hiking associations.) An attempt will be made here to identify species by description, but this is insufficient without a spore print from an actual specimen to confirm the identification.
Recommended references: see page 88. Another book, *Let's Try Mushrooms for Color*, by Miriam Rice, is helpful, but dyers are advised that few species she uses occur in eastern Canada and the northeastern United States.
Parts used: whole mushrooms or puffballs
Processing: chop first, or tear apart with the hands. For maximum results,

use mushrooms which are over-ripe or in a state of deterioration. Soak the mushroom pieces in water to cover from one to three days, stirring the mixture several times daily. The mush will smell strongly, so keep it covered and away from children and animals. The mushroom pieces can be strained off after processing, or left in with the liquor for dyeing by the 'contact' method. (See p 161.)

Colours obtained: this varies with the species used, but most mushrooms and puffballs tested give yellow, gold, tan, bronze, rust, green, and grey. The flesh of some mushrooms, when cut, immediately turns a dramatic shade of bright blue. (The Bigelow reference describes *Boletus frostii* as turning to blue when cut. Groves mentions a *B. subvelutipes* as reacting the same way, but does not list *B. frostii*. Groves gives another species which turns blue when cut, *Gyroporus cyanescens*. *B. frostii* is apparently not dangerous to all who eat it; *B. subvelutipes* is dangerous, and *G. cyanescens* is edible. If inexperienced collectors use the blue factor as a means of identification, they might mistakenly pick a poisonous species. This points out the danger of collecting unknown species.) However, I was unable to obtain blue from these, getting instead a good slate-grey. Rice obtains red, lavender, and magenta from unidentified species of *Cortinarius* (p 39).

Fastness: good to excellent (greys may pale)

How to identify: see above. Dyers can collect other dyestuffs without having to know what, in fact, they are picking. But with mushrooms, it is CRUCIAL to know whether or not a species is poisonous. It cannot be overstated that touching poisonous species with the bare hands is extremely dangerous. All species in the following list are variably edible, and will not cause illness to most people. This means that there are some persons who will react negatively upon eating these species just as some of us get 'hives' from eating certain kinds of fruit. Do not collect species which are described as 'dangerous.' Never take the advice of anyone regarding a mushroom if that person has not actually eaten the species referred to. When collecting, you will notice even deadly (an adjective widely used by mycologists; 'deadly' means just that) varieties such as *Amanita virosa* (destroying angel) appear to have been nibbled at by animals. Species safe for wildlife are not necessarily safe for human consumption, so signs of this 'nibbling' DO NOT indicate that the mushroom is edible. There are other myths as well, notably that all pink-gilled species are safe, which is not true, and that mushrooms which peel easily are edible. This is likewise untrue. There are no instant keys to determine safety: only correct identification gives you the protection you need, and even knowledgeable people can make mistakes. Perhaps one means of reassuring yourself is to first learn to identify two of the deadly amanitas.

POISONOUS MUSHROOMS

Amanita muscaria: known as 'fly agaric,' this common mushroom looks like those pictured in children's books as so-called 'toadstools.' (There is no such thing as a toadstool.) It has a cap from 3 to 8 inches (7–20 cm), which is often bright orange or yellow-orange and covered with white wart-like patches. A beautiful mushroom to look at, *A. muscaria* is common in mixed woods, and I see it far more often than the species *A. virosa*. *Amanita virosa:* destroying angel is so named because of its stately white appearance. It looks pristine and innocent, but is even more deadly than fly agaric. All-white, *A. virosa* has a loose, hanging veil on the stem and a cup at its base, known as the 'universal volva,' two characteristics it shares with *A. muscara*. *A. virosa* is not as common as *A. muscaria*, although they both occur singly and prefer mixed woods as a habitat.

EDIBLE MUSHROOMS

Agaricus campestris	common meadow mushroom commercial species seen in stores	soft yellow (alum); gold (chrome); bright yellow (tin)
Boletis edulis	edible *Boletus*; has pores under cap rather than gills	rust (chrome); warm brown (chrome and tin)
Calvatia gigantea	giant puffball	light yellow (alum); gold (chrome); bright yellow (tin)
Cantharellus cibarius	*Chantarelle* (popular in European cookery)	yellow-orange (alum and chrome); orange-rust (chrome and tin)
Coprinus comatus	shaggy mane (inkies, also back-gilled, are *Coprinus atramentrarius*)	greys
Lactarius deliciosus	delicious lactarius	orange-rust (chrome and tin)
Marasmius oreades	fairy ring, common on lawns	yellow (alum); tan (chrome)
Morchella esculenta	edible morel	warm tan (chrome); dark brown (chrome and iron)

See page 181. People with any food allergies should avoid eating mushrooms.

Availability: mushrooms grow best during damp weather in spring, summer, and fall, although they are most abundant when the days are warm and the nights are cool. (A Dutch friend advised me to pick *Agaricus campestris* in early to mid-September on mornings following a heavy dew. He thinks the best growth is in old cow pastures adjacent to farms that have been worked at least a century. I follow this advice and so always have fresh mushrooms.) Some species grow in the spring only, some in the fall, and a few last from spring until late summer. *Agaricus campestris* prefers pastures, cemeteries, and occasionally lawns; *Boletis edulis* grows in mixed woods; *Calvatia gigantea* is found in fields and pastures; *Cantharellus cibarius* prefers moist locations near conifers; *Coprinus comatus* grows on gravel roadsides and on lawns; *Lactarius deliciosus* prefers mixed woods; *Marasmius oreades* grows on lawns and in woods near hardwood trees; *Morchella esculenta* is found in orchards and pastures in spring.

Mustard

annual wildflower, weed *Sinapis arvensis, Brassica kaber*

There are two scientific names for common mustard: *Sinapis arvensis* and *Brassica kaber* (*Common and Botanical Names of Weeds in Canada* 8 and 31; Frankton and Mulligan 90). Mustard is often confused with a similar plant, wild radish (*Raphanus raphanistrum*). However, mustard flowers are deeper yellow, and mustard flowers throughout the summer. Radish has a longer season of bloom, lasting well into November in some regions. A coarse, hairy-looking plant, mustard may grow as high as 3 feet (.92 m). Look for it in agricultural areas, although it sometimes invades suburbia.

Parts used: whole plant
Processing: chop or tear plant and soak out in water to cover overnight
Colours obtained: pale yellow (alum); pale yellow-green (blue vitriol); tan (chrome); soft grey (iron); medium yellow (tin); gold (alum and tin)
Fastness: excellent
How to identify: *Brassica kaber* (also called charlock, Frankton and Mulligan 90) is similar in overall appearance to *Brassica rapa* (field mustard), but the stem is hairy and the leaves more pointed at the top. Wild radish is similar to both *B. kaber* and *B. rapa*, except that it has a different-shaped leaf. See also Frankton and Mulligan; Gibbons; Martin; Peterson.
Availability: mustards are extremely common throughout Canada and most of the United States. Look for the brassicas in waste places around farms and along rural roads. In suburban areas, brassica often grows near new house sites.

Nasturtium

annual, garden flower *Tropaeolum majus*

A small, many-coloured variety of annual garden flower, the nasturtium is making a comeback in popularity, especially among beginning gardeners, as it is so easy to grow. Available in shades of white, yellow, pink, rose, red, orange, mauve, and violet, some species of newer hybrids grow to 2 feet (.61 m) in height but the dwarf types are more common.

Parts used: blooms (fresh or as they fade)
Processing: as for flowers
Colours obtained: faded blooms, mixed shades: yellow-beige (alum); yellow-tan to gold (chrome); grey (iron); bright gold (tin); yellow-green (blue vitriol)
Fastness: good
How to identify: see photographs in DO, MC, ST, and V. Available as seed from these and local nurseries and as bedding plants

Nettle

weed *Urtica dioica*

Although wild food enthusiasts consider the stinging nettle fine food if harvested early in the season, most dyers know that it is a frustrating dyestuff to collect even when wearing gloves. *Urtica dioica* is the nettle most often used by dyers, but a plant called purple or dead nettle (Peterson 280, *Lamium purpureum*) is just as useful. It looks much like nettle but in the fall the entire plant turns reddish-purple. It is not as difficult to collect as the stinging nettle and when used fresh the two plants give almost identical colours.

Parts used: whole plants
Processing: chop first and soak out in water to cover overnight or for a day or two
Colours obtained: yellow-green (alum); strong yellow-green (blue vitriol); gold (chrome); greyish-green (iron); chartreuse (tin). *L. purpureum* collected for dyeing after it has turned reddish-purple (mid-September) gives: rosy-brown (chrome); warm tan (alum); pinkish-grey (iron)
Fastness: excellent for shades from fresh nettle or purple nettle; good for colours from purple nettle in the fall
How to identify: Cunningham; Martin; Peterson; Stewart and Kronoff
Availability: *U. dioica* is more common than *L. purpureum* in Nova Scotia (Roland and Smith 344, 601), but both seem often to be quite localized in

their habitat. Both grow on waste ground, near gardens, and along roadsides. I collect *L. purpureum* near new house sites. Do not touch stinging nettle with bare hands.
Special notes: Stewart and Kronoff write that a green dye was made from the whole plant and a yellow dye from the roots (p 22).

Oak

hardwood and ornamental tree *Quercus*

There are several varieties of oak common to the eastern portion of the continent, including *Quercus borealis* (red oak; Saunders 57), *Q. alba* (white oak), and *Q. rubra* (designated red oak by Hosie 192 and Petrides 218. Neither lists *Q. borealis*. Roland and Smith have a listing of *Q. borealis Michx. f.*, p 341, which they call 'red oak'). Different authors use various species names depending upon their own research; plants are often reclassified. The introduced English oak is *Q. robur*. The traditional dye oak, black oak, is *Q. velutina* and is limited in Canada to the Niagara Peninsula area of Ontario, with scattered growth north to Belleville and Trenton (Hosie 194. Furry and Viemont write that the American habitat includes Pennsylvania, Georgia, and the Carolinas, p 11). This is the oak traditionally used by dyers in past centuries as a fine source of fast yellow on wool and cotton (Adrosko 47). Called quercitron, the dye was prepared after making an extract from the inner bark of *Q. velutina*. Hosie notes that black ink can be made from galls of pin oak, *Q. palustris*, by adding iron filings to water in which the galls are soaking out (p 196). Black can be obtained by adding various mordants to oak bark, in addition to the iron filings or an old horseshoe, but the black is really more of a slate-grey.

Parts used: acorns (see p 89); fresh or fall leaves; bark
Processing: according to type. Quercitron is made by removing the inner from the outer bark of *Q. velutina*, drying it, and then pulverizing it into a powder. The bark from other oaks is removed with an axe or hatchet (felled trees only) and then allowed to soak out in water until it is soft enough to break up into small pieces. Allow these to soak further. Proceed with the dyeing when the water in which the bark is soaking has taken on a reddish-rust colour. Because of the tannin in bark, the temperature of the dyebath should be kept below a simmer if colours other than brown are desired.
Colours obtained: see acorns; fresh leaves: yellow (alum); tan (blue vitriol); gold (chrome). Fall leaves: tan (alum); medium brown (alum and chrome); rusty tan (tin). Bark from *Q. borealis*: gold (alum); bright

gold (chrome); strong yellow (tin); grey (iron filings, horshoes, and blue vitriol); reddish-brown (vinegar and tin). Oak bark combined with sumac fruiting cones and nails yielded a strong dark grey. (Furry and Viemont, pp 11–12, and Davidson, p 15, list colours from many *Quercus* species.)
Fastness: fresh leaves: excellent. Barks: excellent. Fall leaves and acorns are quite fast but all the browns are likely to change and darken with time.
How to identify: MC and SH offer different varieties but neither lists *Q. velutina* (black oak). However, quercitron is available as a dyestuff from dye suppliers.

Olive

ornamental tree *Elaeagnus angustifolia*

Planted as an ornamental, the Russian olive is a delicate tree with lovely silver-green foliage and interesting grey bark. It is not a common tree, but when planted as nursery stock will grow rapidly and reward the dyer who uses its leaves by producing more than enough foliage for ornamentation as well as dyeing.

Parts used: fresh foliage
Processing: as for leaves
Colours obtained: a clear and rich yellow is obtained using alum; tin gives a bright gold, and chrome a warm, deep gold-olive; iron gives a yellow-grey and vinegar a warm beige
Fastness: excellent
How to identify: Hosie; Knobel; Petrides
Availability: as nursery stock, from MC and SH and some local nurseries

Onion

vegetable *Allium cepa*

Read pages 58–60.

An onion skin dyebath is usually the first one a beginning dyer makes because the dyestuff is readily available and the results are known to be almost foolproof. At one time or another, most of us have dyed Easter eggs using onion, and many oldtimers will tell you that they remember their mothers or grandmothers using onion skins to dye everything from flour sacks to linens. It is unfortunate that many contemporary dyers pass off an onion skin bath as being too ordinary to bother using. After all, most dye workshops have an onion pot on the stove, and the results are rather predictable if few mordants are used. I think that no other dyestuff offers quite the potential of onion when all possible variations are put into play, including using a wide variety of mordants, top-dyeing, and dipping non-white fibre in the bath. I have dyed everything from bones to feathers to shells and hair with onion, always with exciting results. Pale onion skins processed in an enamel pot produce a colour totally different from pale skins in an iron pot. Dark-skinned home-grown onions give rich shades that almost defy description: burnt orange, rust, rust-orange, rich dark brown, and olive-brown. Fermented skins allowed to sit for several weeks give exciting colours, as do those used to dye grey and brown fibres. Onions can be added to bark baths to lessen the brown effect and increase the yellow. Fibres topped or bottomed with onion (before or after indigo, madder, etc.) are very beautiful, as are yarns dyed in a bath of onion combined with, say, mullein. Onion skins can be used in this way to 'stretch' a dyestuff. Cotton and linen are easily dyed with onion skins, as are such materials as cotton eyelet and other laces. Even T-shirts can be dipped in the onion bath and effectively tie-dyed. What other dyestuff is as available and as safe to use even with small children? (When dyeing with children present, use only household mordants such as salt and vinegar.)

Parts used: skins from any variety of onion (home-grown onions give darker colours); white or yellow-skinned market varieties; skins and/or whole onions used chopped in a single bath. (At a Halifax workshop in September 1974, Edna Blackburn of Caledon Hills, Ontario, suggested using 3 lbs (1.5 k) of onions, chopped, to 1 lb (453 g) of fibre.)
Processing: be flexible, and try as many different methods as you can think of. The skins can be soaked, left to ferment, and used in a weak, medium, or strong bath. (See pp 23 and 76.) You do not need mordants, but their use will extend the colour possibilities.
Colours obtained: the following list is only a guide. The bath was medium, made from dark-skinned home-grown onions. Skins were not pre-soaked and all yarns were pre-mordanted and left damp until the dyeing.

colour	mordant
yellow	alum
bright orange	tin
bright yellow-orange	alum, bloomed in tin
rust	chrome
rich brown	chrome and tin
olive-green	iron
khaki	blue vitriol

To achieve the softest yellows (lemon), use only a few skins and process the fibre in the bath for less than half an hour. No mordant is required. For rust with an alum mordant, use very dark-skinned onions in a medium or strong bath, and pre-soak them for 24 hours. For old gold, use light-skinned onions in a weak bath with alum and chrome. Vinegar will give dark soft yellows and salt gives a lovely brass if processed in an iron pot. The colours obtained depend upon the type of skins used, the pot, the water, and the mordants.

Fastness: mordanted shades are very fast; unmordanted colours are fast initially, but may fade after repeated washing and exposure to light.
Availability: as seed, from DO, MC, ST, and V and local suppliers: as 'sets' from local nurseries. Look for skins at vegetable packing plants, restaurants, and institutions.
Special notes: both Robertson (p 23) and Grae (p 182) report obtaining a lime or olive-green from red onion skins, using a tin mordant. Kierstead includes a New Brunswick recipe in her book which calls for adding alum and cayenne pepper to the onion bath to produce brilliant orange (*Natural Dyes* 52).

Pansy

annual garden flower *Viola*, various spp.

The bright-coloured velvety petals of the pansy make it a favourite everywhere. New hybrids have been developed in recent years to greatly expand the available colour range and height of this flower. Blooms are now quite large in some species, and if you haven't enough, they can be combined in a dyebath with other flowers of a similar colour. The flowers can be picked up as they fade, and kept in a water-filled container until you have enough for a bath. Keep this mixture covered, in a warm spot. It will smell, but reward you with good shades.

Parts used: fresh or faded flowers
Processing: as for fresh flowers, or see above for faded blooms
Colours obtained: maroon, blue, and purple flowers: tan (vinegar); pinkish-beige (alum); gold (chrome); grey (iron). Mixed flowers, faded, allowed to soak in water to cover for two weeks: soft yellow (alum); brilliant yellow (tin); warm bright gold (chrome); lovely soft grey (iron); warm tan (blue vitriol). Grae reports a blue-green with alum but advises that pansy shades fade and are most fast if subjected to a cupric sulphate or copper rinse (p 128).

Fastness: good. The shades obtained from fermented flowers allowed to soak two weeks were quite fast to light.
How to identify: nursery catalogues
Availability: CK, DO, MC, ST, and V as seed; available at most local nurseries as bedding plants

Parsley

herb, annual or biennial *Petrosilinum* spp.

Some gardeners treat parsley as an annual, planting it from seed each spring. But I find the seed difficult to germinate, so I treat mine as a biennial, letting it go to seed the second year. Erichsen-Brown (*Herbs in Ontario* 20) writes that it takes the seed from 3 to 8 weeks to germinate, and this frustrates many novice gardeners. Those who use parsley just as a garnish are missing a real treat! If you have enough, then save some for the dyepot or collect what is discarded on dinner plates at parties, church socials, and teas. As Erichsen-Brown points out, parsley contains more vitamin C than oranges (p 21), yet we throw it away as if it were a frivolous food!

Parts used: leaves and stems (try to locate and use discarded parsley if possible. Wash it off first if other foodstuffs are mixed in with it).
Processing: as for fresh leaves
Colours obtained: soft yellow (alum); bright gold (chrome); bright yellow-green (tin); yellow-grey (iron)
Fastness: good
How to identify: see *Herbs of Ontario* and *Growing Savory Herbs*
Availability: see above references for details on growing parsley from seed. As seed, from CK, DO, MC, ST, and V; from local seed suppliers; some nurseries which specialize in vegetables sell parsley as a bedding plant. It is usually in great demand, even as seed, and many suppliers are out of parsley seed by February or March. Collect it from institutions and organizations holding special suppers and teas.

Parsnip

vegetable *Pastinaca*

Parsnip greens, or the leafy tops, make an excellent dye. Parsnip is a yellowish vegetable with a long edible root that looks like a thick carrot.

The leafy green top is taller than a carrot top and not as delicate and fern-like in appearance. Ask neighbours who grow parsnips to save the tops for you, or collect them from farm produce stands and supermarkets.

Parts used: green leafy tops
Processing: as for fresh leaves
Colours obtained: weak bath: lovely clear yellow (tin); medium bath: dull yellow (alum); gold (chrome); tan (blue vitriol)
Fastness: excellent
How to identify: illustrated in seed catalogues; ask locally
Availability: as seed from DO, MC, ST, and V. As seed from local suppliers

Pea

vegetable, wild seaside plant *Pisum sativum, Lathyrus japonicus*

Both the domestic and wild pea are useful to the dyer. Although domestic pea (*P. sativum*) pods are disappointing, because so many are needed to give even a very pale colour (beige, with vinegar), the vine gives some colour in a medium or strong bath. But the wild beach pea, *L. japonicus*, is not only very plentiful along coastal regions but a better performer in the dyepot as well. The tiny peas collected from the mature pods of the beach pea in August are an excellent wild food.

Parts used: domestic pea: vines. Wild beach pea: whole plant
Processing: using either type, chop or tear up the vine or plant and soak it out in water to cover for a day before proceeding with the dyeing
Colours obtained: domestic pea, pods: beige (vinegar); light tan, very pale (chrome); tan (blue vitriol). Domestic pea, vine: yellow (alum); bright yellow (tin). Beach pea, whole plant: yellow (alum); gold (tin); yellow-green (blue vitriol); dark gold (chrome and alum); yellow-grey (iron); taupe (chrome and iron)
Fastness: domestic pea pods and vines, good. Beach pea, excellent
How to identify: MacLeod and MacDonald; Peterson; Stewart and Kronoff for beach pea
Availability: P. sativum is available as seed from DO, MC, ST, and V, as well as from local suppliers. Look for *L. japonicus* growing along beaches on sand dunes and in fields near beaches.
Special notes: the garden flower, sweet pea, is *L. odoratus*. Although I found no reference to use of this as a dyeplant in any source, there is no reason not to try it. Perhaps the fact that it is not used is appropriate motivation.

Pear

ornamental, fruit tree *Pyrus communes*

The pear is usually grown for its fruit, but because the spring blossoms are so decorative, some species are prized as ornamentals. As is the case with apple, peach, and cherry trees, pear is now available in dwarf sizes, which means even city-dwellers can enjoy fresh fruit off their own tree. The leaves are used for dyeing when picked fresh; they may be combined with leaves from other fruit trees, such as those listed above. Pick only a few leaves from each branch, unless the tree is large.

Parts used: fresh leaves, alone or combined with leaves from other fruit trees; fresh peeled skins (collected during canning and preserving season)
Processing: process leaves as for fresh leaves. As skins are collected, place them in a water-filled container to which you have added 1 tsp (5 ml) of baking soda. Use right away, or when you have saved up a sufficient amount. The colours obtained are much like those from banana peels (see p 97) but paler.
Colours obtained: leaves: soft lemon-yellow (alum); gold (chrome); bright yellow-orange (alum and tin); beige (blue vitriol); grey-tan (blue vitriol and iron). Robertson gets soft yellow from the bark of pear (p 22).
Fastness: excellent for leaves; skins not tested for light fastness
How to identify: Knobel; Petrides
Availability: from MC, SH, and local nurseries as nursery stock. Look for old pear trees adjacent to deserted farms in rural areas.

Pearly-Everlasting

perennial wildflower *Anaphalis margaritacea*

As delightful as its name sounds is the appearance of this charming wildflower that many people bring indoors for winter decoration. The flat-white flowering head is similar in shape to that of the flat-topped goldenrod (see p 148) but it is woolly-looking and smaller in size. The stem of the pearly-everlasting is whitish and flannel-looking, rather like mullein (see p 179). There are several other so-called everlastings, which Roland and Smith refer to as *Antennaria* (p 675). What Peterson calls sweet everlasting, or catfoot, is *Gnaphalium obtusifolium* (p 90).

Parts used: flowering heads or whole plant
Processing: as for flowers and whole plants

Colours obtained: whole plant: strong yellow (alum); chartreuse (alum bloomed in tin); gold (chrome); yellow-green (blue vitriol and iron)
Fastness: excellent
How to identify: Cunningham; Roland and Smith; Peterson
Availability: look for pearly-everlasting from August on, growing in poor soil, often adjacent to highways

Peony

garden flower, perennial *Paeonia lactiflora*

Peonies are full-bloomed perennials with large flowering heads which may look like roses to the novice gardener. Available in a wide range of colours, the blooms are often 4 or 5 inches (7.5–12.5 cm) across. Peonies bloom fairly early, usually in June, and last for several weeks. As each flower has so many petals, collecting only a few faded blooms will provide the dyer with enough for a dyebath. As with all lovely flowers, the blooms should be used for dyeing only after they have served their purpose as ornamentation. Different shades can be mixed together in a single dyebath or blooms of one colour saved until enough have been collected for a bath.

Parts used: faded blooms
Processing: as with all flowers used after they have faded (daffodils, roses, tulips, and so on; see p 70). Peonies can be soaked in water and left to ferment in a warm place for two or more weeks. The smell is outrageous, but the colours produced by this extended period of soaking are worth it.
Colours obtained: mixed colours of blooms: yellow (alum); sharp yellow-gold (tin); strong medium gold (chrome); bronze (blue vitriol); orange-tan (iron and chrome)
Fastness: excellent
How to identify: see photographs in nursery catalogues; ask locally
Availability: as seed from DO, MC, ST, and V. As stock from CK and SH. Available from local nurseries

Petunia

garden flower, annual *Petunia*

Marigolds, geraniums, and petunias are probably the most commonly planted garden annuals. And with good reason. If the blooms are picked off as they fade, the petunia will continue to flower until very late in the season. New hybrids have been introduced to expand both the size and colour range of the blooms. There is an amazing variety of shades available, including one that is red and white. Cruikshanks catalogue has a yellow petunia available as seed.

Parts used: faded blooms
Processing: treat by soaking in water to cover as for other faded flowers (see processing of peony, p 192)
Colours obtained: mixed shades: yellow-tan (alum); bronze (blue vitriol); gold (chrome); yellow-green (iron); sharp gold (tin)
Fastness: good, but most dye references do not list petunia because it is suspected of not being fast to light. Grae uses petunia only in combination with other flowers such as marigolds (p 129). My petunia baths were processed using water that was quite hard, and possibly this increased the fastness of my colours.
How to identify: seed catalogues; ask locally
Availability: as seed, from CK, DO, MC, ST, and V; local seed suppliers. As bedding plants, petunias are available from even the smallest nursery.
Special notes: New Englanders are partial to planting a bed of purple and white petunias in front of tall, scarlet geraniums. This produces what can only be described as visual delight, especially when the bed is in front of a dark-stained house or building.

Pigweed

See lamb's quarters, page 162.

Pine

coniferous tree *Pinus*

Unlike other conifers, pine is easily distinguishable by its long needles and lovely aroma. The needles of white pine (*P. strobus*) and red pine (*P. resinosa*) are from 3 to 5 inches (7.6–12.7 cm) in length, while those of fir, hemlock, spruce, and larch are noticeably shorter. Because of its great value as timber, dyers wishing to try the bark are advised to visit sawmills. Sawmills are a source of bark from other lumber trees as well, including spruce, fir and, if you're lucky, birch and maple. The noise is deafening! Do not take children or pets to milis. Have large containers for the bark, which may be in enormous slabs.

Parts used: needles, cones, bark
Processing: treat all as for bark, soaking out first in water to cover for one or two days. Add salt to the needle bath to help draw out the pigment.
Colours obtained: fresh needles: yellow-beige (alum); brown needles (off the ground): tan (vinegar); gold (chrome); brown (chrome and iron); cones: warm tan (alum and chrome); bark: warm tan with a reddish tinge (alum, chrome); pinkish-red (alum, bloomed in tin). This last shade was odd: decidedly reddish but not at all tan.
Fastness: good for all; excellent for shades from bark
How to identify: Hosie; Knobel; Petrides; Saunders; Sherk and Buckley. Most nursery catalogues have photographs of trees.
Availability: throughout Maine and New Hampshire are the very finest native pines, growing along roadways in urban as well as rural districts. Pure pine stands are, unfortunately, not common in Nova Scotia, but can be found in sandy parts of Annapolis County. *P. strobus* grows in Cape Breton, on Prince Edward Island, Newfoundland, and through New Brunswick west to the Ontario/Manitoba border (habitat as given by Hosie 36). As nursery stock from MC and SH, and local nurseries
Special notes: Grae obtains an olive-green from pine needles processed in an iron pot (p 171). Thurston lists a reddish-yellow from a cone bath (p 24). But Saunders scores the highest with this gem of information borrowed from an English writer by the name of Josselyn: apparently water in which the immature cones are soaked will remove wrinkles from the face when applied with a face cloth (Saunders 16).

Plantain

weed *Plantage major*

Plantain is a very common weed, disliked because it grows in cracks in cement, gardens, lawns, and indeed, wherever a weed is not welcome. A very persistent and hardy plant, plantain can be recognized by its ribbed leaves which lie almost flat on the ground, and single flower spike with inconspicuous green flowers that later turn to mealy-looking seeds. It flourishes from June through summer and into fall. While many so-called weeds are, to some of us, lovely and welcome wildflowers, plantain really can be described as 'weedy.' However, its young leaves are considered a very fine wild food. Indeed, one reference on the subject refers to *P. maritima* as being used in Nova Scotia as a summer vegetable (Stewart and Kronoff 26) but MacLeod and MacDonald list only *p. major* and *P. lanceolata* (p 66).

Parts used: the whole plant; use the root as well if you are digging them up to be rid of them for good
Processing: as plantain leaves are rubbery in texture they must be torn up well, and then the whole plant soaked in water to cover for one or two days
Colours obtained: from new leaves only, picked in early June: soft green (alum); gold-bronze (blue vitriol); yellow-green (chrome); grey (iron); grey-green (tin)
Fastness: excellent
How to identify: Cunningham; Frankton and Mulligan; Gibbons; MacLeod and MacDonald; Martin; Peterson; Stewart and Kronoff
Availability: plantain is an abundant weed throughout Canada and the north, central, and west United States. It seems to prefer poor soil and often grows up through cracks in pavement. Plantain edges roads, sidewalks, and frequently locates itself right beside the back door, hence the term 'dooryard' weed.

Plum

ornamental and fruit tree *Prunus*

As is the case with apple, cherry, pear, and peach, the leaves and bark of the plum yield excellent dyes (*Prunus* is the genus for plum, peach, cherry, almond, and many other fruits. Because there is little chance of confusion it is sufficient to call a plum 'plum' and a peach 'peach.'). The fruit may also be used, especially if it is diseased or otherwise inedible. Dyers who live in areas where plum grows wild (*P. nigra*, Canada plum; and *P. americana*, wild plum) may wish to try these fruits for dyes. Leechman gives a recipe for the bark of *P. nigra*, and quotes the British astrologer and physician, Culpeper, who wrote that all plums are under Venus and, like women, some are better than others (p 51).

Parts used: leaves, bark, fruit, and/or their skins
Processing: leaves and bark are processed according to type. The fruit and skins may be combined with peelings from other fruits to make up a bath. The residue left over after jam-making can be used for a dyebath by adding to it an amount of water sufficient to make the mixture 'thin.' Soak the fibre to be dyed in the fruit and water prior to dyeing for at least a day. After processing, wash the dyed fibre well to remove all traces of fruit pulp.
Colours obtained: fresh leaves (from domestic dwarf plum): yellow (alum); bright yellow-orange (tin); rust (alum and chrome); gold (chrome). Bark

(domestic): reddish-brown (alum and chrome); taupe (alum saddened in iron). Fruit, mixed in with cherries: soft pink (alum); rose (chrome); pinkish-grey (iron). See special notes below.
Fastness: excellent for all but those shades obtained using the fruit. They were only fast to light for a few days, and faded noticeably when washed.
How to identify: Hosie; Knobel; Petrides; Sherk and Buckley
Availability: Hosie (pp 248, 250) and Petrides (pp 194, 195) give different habitats for both *P. nigra* and *P. americana*, but of the two, *P. Nigra* (Canada plum) has the more extensive range, occurring from the Maritimes west to Manitoba. Roland and Smith (p 469) list *P. nigra* as occurring in the Annapolis Valley of Nova Scotia as an introduced species that has escaped to the wild. From MC and SH as nursery stock, and from most local nurseries
Special notes: Thresh and Grae obtain olive-green from plum, Grae from the fruit, with ammonia (p 183) and Thresh from the leaves of Japanese flowering plum (p 22). What is interesting about the Thresh recipe is that, after dyeing, the fibre was then exposed to strong sunlight to develop the colour further (see p 172).

Poplar

hardwood tree *Populus nigra*, var. *italica*

See page 94 for aspen (*P. tremuloides*). *P. nigra* var. *italica* is the Lombardy poplar which, because of its tall, columnar shape is frequently planted as an ornament. Although its leaves are acknowledged by most dye references to give a fine yellow, the leaves of the aspen produce the same shade. The Lombardy poplar was once very common. Extremely fast-growing, it is a desirable tree to use when planting a boundary between properties. *P. nigra* 'Thevestina,' Theve's poplar, is very similar to the Lombardy, but hardier.

Parts used: leaves
Processing: according to type
Colours obtained: yellow (alum); bright yellow-orange (tin); bright gold (chrome); rust (tin and chrome); brown (chrome and iron); brown (blue vitriol)
Fastness: excellent
How to identify: Hosie; Knobel; Petrides; Saunders
Availability: from MC and SH as nursery stock. Available at local nurseries
Special notes: the poplar referred to by both Davidson (p 17) and the Krochmals (p 95) is *Liriodendron tulipifera*, or the tulip tree.

Poppy

perennial garden flower *Papaver orientale*

Poppies are tall-growing perennials in shades of white, yellow, pink, salmon, orange, and red. Icelandic poppies (*P. nudicaule*) bloom in June but the Orientals (*P. orientale*) last until July. If you have too few poppy blooms, mix them in with lilies, petunias, or pansies of a similar colour.

Parts used: fresh or faded blooms, alone or in combination with other flowers
Processing: as for flowers
Colours obtained: from red-orange Icelandic poppies: warm beige (alum); yellow-gold (tin); dark gold (chrome); pinkish-tan (blue vitriol); warm grey (iron). Grae reports obtaining a mustard yellow from seed pods (p 130).
Fastness: excellent
How to identify: Herwig; nursery catalogues have photographs. Icelandic poppies are among the earliest red-blooming perennials and grow from 2 to 3 feet (.61–.92 m) in height.
Availability: as seed from CK, DO, ST, and V; as nursery stock from MC and SH. CK offers dwarf types (*P. alpinum*) and annuals (*P. glaucum*)

Primrose

perennial garden flower *Primula*

See evening primrose, p 140.

The primrose is probably more popular as a garden flower in England than in Canada and the United States. There are many varieties. *P. vulgaris* is called the English yellow primrose, while *P. polyantha* has blooms of many colours. *P. juliae* flowers as early as May, and most primroses will bloom an entire season if the seeds are removed as they form.

Parts used: flowers
Processing: as for flowers
Colours obtained: mixed shades: pale yellow (alum); bronze (blue vitriol); warm tan (chrome); bright yellow (tin)
Fastness: excellent
How to identify: Herwig; nursery catalogues have photographs
Availability: as seed, from CK, DO, ST, and V; as nursery stock from MC and SH and local suppliers

Privet

deciduous shrub *Ligustrum ibolium*

Although *L. ibolium* is probably the most popular privet used as shrubbery and for hedges, there are many other varieties available such as *L. obtusifolium*, which has berries in the fall. Grae reports disappointing results from the berries of an unspecified *Ligustrum* (p 172), but Robertson uses the fruit of *L. vulgare* for pinks, blues, and purples (p 67). She does not mention whether the shades are fast. However, the leaves of all species of *Ligustrum* are acknowledged as a fine source of fast yellow.

Parts used: freshly picked leaves, free from twigs; berries (see above)
Processing: soak out in water to cover several hours before dyeing
Colours obtained: leaves: bright, clear yellow (alum); yellow-green to chartreuse (tin); gold (chrome); bronze (blue vitriol and iron); yellow-grey (iron)
Fastness: excellent for shades from leaves; berries were not tested
How to identify: Knobel; Petrides; Sherk and Buckley; MC and SH have photographs
Availability: MC and SH offer various species. Privet is offered as nursery stock from most local nurseries.

Purslane

weed, wild food *Portacula oleracea*

Purslane is a nuisance to home gardeners but a favourite of wild food enthusiasts. As a child I remember chewing on the tender leaves of this low-growing annual, which would often spread out between the rows in the vegetable garden. Purslane stems are thick and reddish. They are somewhat rubbery in texture, so the plant should be well torn up before soaking it out for a dyebath. As a food, the young leaves may be eaten raw or tossed in with a salad of other greens. One of the most interesting aspects of purslane which makes me fond of the plant is the fascinating reference to it I found in *A Heritage of Canadian Handicrafts*. (See p 129.) Apparently New Brunswick Indians used purslane with alum for a bright blue. This is doubly interesting as no other dye reference I have consulted lists purslane, not even the indefatigable Ida Grae. I give her credit for experimenting with an incredible range of plants. If any dyer has knowledge of the blue from purslane, I would appreciate hearing from them. I gave up trying to obtain this colour, as my results were nowhere close to blue at any time. Still, the plant certainly bears investigation. Some indigenous skills have been almost entirely lost, and it is quite possible that the information documented by the editor, Gordon Greene, died with the person who gave it to him. One may be inclined to doubt such unusual results from a plant such as purslane, but remember that native peoples had a vast lore of plant-related information, most of it handed down in the oral tradition. Sadly, much has been lost.

Parts used: whole plant, torn apart and soaked out in water to cover for two or three days before processing
Processing: as above
Colours obtained: yellow (alum); beige (vinegar); tan (blue vitriol); yellow-green (alum and tin); gold-brown (chrome). See above for blue.
Fastness: good
How to identify: Frankton and Mulligan; Gibbons; MacLeod and MacDonald; Martin; Peterson; Stewart and Kronoff
Availability: purslane is extremely common throughout Canada and much of the United States, particularly near gardens, lawns, and cultivated areas. It has a spread of up to 2 or 3 feet (.61–.92 m) and hugs the ground very closely.

Queen Anne's Lace

See wild carrot page 113.

Radish

vegetable, weed *Raphanus*

See page 183. The wild radish is *Raphanus raphanistrum*, and the domestic garden vegetable *Raphanus sativus*. Dyers who keep accurate documentation of their dyeing experiments should learn to distinguish between *Raphanus* and *Sinapis*. The radish leaf is more lobed than that of mustard. Radish flowers are small and may be light yellow or even white. Mustard flowers are always yellow and darker in colour than those of the radish. The domestic radish has non-yellow flowers, and many home gardeners find it goes to seed long before they have used all of the vegetable they planted.

Parts used: wild radish, whole plant. Domestic varieties, green tops
Processing: according to type. A strong bath is recommended for domestic tops.
Colours obtained: soft yellow (alum and tin); bright yellow-gold (chrome and tin); beige-tan (blue vitriol); soft grey (iron). Domestic radish tops give medium green with iron in a strong bath.

Fastness: good
How to identify: Frankton and Mulligan (Frankton and Mulligan 88–90, show both mustard and radish on subsequent pages, so one can easily compare them); Martin; Peterson
Availability: wild radish is an extremely common weed, especially around new house sites, agricultural areas, and subdivisions. It grows on poor soil that is sandy or dry. As seed, from CK, DO, MC, ST, and V. Radish is easily grown by even the novice gardener.
Special notes: another way to differentiate between radish and mustard is to learn their seasons. Wild radish blooms very late in the fall, often until November, while mustard is through flowering by late summer.

Ragwort

wildflower, weed *Senecio*, various spp.

Many dyers confuse ragwort (*Senecio*) with tansy (*Tanacetum*), which is not surprising when you realize that common ragwort, *S. jacobaea*, is called tansy ragwort (Frankton and Mulligan 180) and sticking Willie. The latter is surely a very descriptive vernacular name, but does not imply what a dangerous weed this is to livestock. All the ragworts are excellent sources of dye, including *S. aureus* (golden ragwort) and *S. robbinsii* (swamp ragwort). The ragworts grow to 3 feet in height (.92 m) and have yellow flowers. The blooms of *S. jacobaea* form a flat-topped flowering head which rather resembles tansy, but the latter is a plant with a very lovely aroma.

Parts used: whole plant or flowers alone
Processing: as for whole plants or flowers
Colours obtained: yellow (alum); bright yellow (tin); bright gold (chrome); grey-beige (iron); yellow-tan (blue vitriol). Using just the flowers gives shades that are yellow, yellow-green, or chartreuse.
Fastness: excellent
How to identify: Frankton and Mulligan; Peterson
Availability: common throughout Canada and the northeastern and north-central United States. Ragwort grows in fields, pastures, and along roadsides.
Special notes: The ragwort Thurston refers to (p 25) as growing widely in England is *S. jacobaea*. Robertson (p 46) writes that *S. jacobaea* is also known as 'ragweed,' but Frankton and Mulligan call *Ambrosia artemisifolia* by that common name and not *S. jacobaea* (p 116). Peterson (p 176) advises readers that ragwort and ragweed are often confused with each other. It is ragweed (*A. artemisifolia*) that is often blamed as a cause of

hay fever. It has tiny green flowers and does not resemble ragwort except in name.

Raspberry

wild and cultivated fruit *Rubus idaeus, Rubus strigosus*

That part of the wild or cultivated raspberry (see blackberry, p 105) which yields the most interesting results in the dyepot is the leaf, not the berry. The canes, too, make an excellent dyestuff. Unless it is the residue left from jelly-making, I cannot personally justify the use of the fresh fruit for a dye. This is a personal opinion. My family will happily eat fresh berries at each meal. Berries give colours that are pinkish at first, but turn to pink-tan or taupe as soon as the water in the dyebath reaches a simmer. The colour obtained from raspberries may only be fast if sugar is added (see p 75). Raspberry and blackberry grow in such abundance that taking a few canes will not matter. Both reproduce readily. There are many species of *Rubus*, including those that go by the following common names: red raspberry; wild raspberry; black raspberry; dewberry; and blackberry. Roses also belong to the genus. Some references list *R. idaeus* and *R. strigosus* interchangeably (Cunningham 68) but Roland and Smith (p 456) consider *R. idaeus* as the cultivated variety (although often escaped) and *R. strigosus* as wild raspberry.

Parts used: fresh or frost-bitten leaves; new shoots or mature canes; berries if desired
Processing: see blackberry, page 105
Colours obtained: colours obtained are almost identical to those from blackberry, but the fruits, in a strong bath with sugar and flour, will give a rose-tan or a rose-grey with vinegar. Fruits steeped in water will give a pinkish shade to wool fibres pre-mordanted with alum, but the colour is not fast.
Fastness: highly variable with berries; colours from leaves and shoots are excellent
How to identify: Cunningham; Gibbons; Knobel; MacLeod and MacDonald; Petrides; Stewart and Kronoff
Availability: as nursery stock from MC and SH and local nurseries. Wild canes are easily transplanted to similar soil.
Special notes: just as Thurston obtained a black from new blackberry shoots with an iron mordant (see p 106), the Krochmals list raspberry shoots as a source of grey and black (p 233). The mordant used, for a pound (453 g) of fibre is 3 Tbsp (45 ml) of iron, an amount which would render the fibre extremely harsh to the touch. The cooking period, at the

boil, was two hours. I tried this. My sample was slate grey but quite unusable owing to its unpleasant crimpy feel.

Rhododendron

flowering evergreen shrub *Rhododendron*

Rhododendrons are evergreen shrubs prized as ornamentals. It is the leaves of the plant which are used for dyeing, however, and not the blooms. Unless you have rhododendrons of your own, don't expect a gardener who has two or three shrubs to give you leaves. Find another source. I solved this problem for myself by visiting the local Department of Agriculture research station to make inquiries. I was told that an entire field of rhododendrons was being removed, due to winter kill. I was free to pick literally bags of leaves. Walter Ostrom, a potter who lives at Indian Harbour, NS, suggested another source to me, and that is *R. canadense*. Commonly called *Rhodora* (Cunningham 90; Petrides calls *R. canadense* 'Rhodora azalea' 284) this small, woody shrub has magenta flowers which bloom in April and May, appearing before the leaves. It is deciduous, unlike the rhododendron. A similar evergreen shrub is laurel, *Kalmia angustifolium*. This low perennial is very common in the northeast, especially on Prince Edward Island. It gives results similar to rhododendron.

Parts used: leaves
Processing: as for fresh leaves. Rhododendron leaves are somewhat rubbery and should be torn up before making the dyebath. Just doing this will stain your hands yellow.
Colours obtained: Leaves, picked in early June: beige (vinegar); bright yellow (alum); gold (blue vitriol); bright rust (chrome); bright orange (tin); khaki to olive-green (iron)
Fastness: excellent
How to identify: Cunningham; Knobel; Petrides; Sherk and Buckley
Availability: as nursery stock from SH and local nurseries specializing in ornamental shrubbery. Dyers can visit local parks for leaves from winter-killed rhododendrons, or try the wild species suggested, or *Kalmia*. Farmers are always eager to be rid of laurel. Its common name is sheepkill (or lambkill). The leaves contain andromedatoxin.

Rhubarb

vegetable *Rheum*

Although rhubarb is treated by the seed catalogues as a 'vegetable,' it is used in the kitchen as a fruit, to make superb pies, puddings, jams, and jellies. Only one book, *Natural Plant Dyeing* (p 22), refers to rhubarb as the wonderful dye source it is. As it usually takes at least a dozen or more good-sized stalks to make a pie, there are normally enough leaves on hand to make a dyebath. Never store these where children can get at them as they are poisonous.

Parts used: fresh leaves
Processing: tear the leaves up, washing your hands well afterwards. Soak them out in water to cover for several hours.
Colours obtained: a tin mordant gives a brilliant orange, alum an orange-yellow, and chrome a rust. (Students have failed to get these results, so use a very strong bath for the orange.) See special notes for colours from the stalks.
Fastness: excellent
How to identify: rhubarb is a standard feature in most rural gardens. The plant has very large green leaves with wavy edges and grows in a clump. The stalks are pink, green, or red. Clumps are easily transplanted in early spring or fall.
Availability: as plants from DO, MC, and SH. From most local nurseries
Special notes: the Gerbers, in *Natural Plant Dyeing*, give lavender as the colour obtained from rhubarb stalks (p 22). The mordants were oxalic acid and tin, but my own results did not confirm this. Tin used alone with the stalks produced a pink-rose, but the colour was not fast.

Rose

cultivated and wild flowering shrub *Rosa*

There are many varieties of rose, both cultivated and wild. The petals of all domestic varieties, such as floribundas, multifloras, and hybrid teas can be used for a dye. Wild rose petals also give interesting colours, as do the stems (canes) and leaves of wild species. Most gardeners will give faded blooms from their ornamental roses and stems, if they are pruning.

Parts used: cultivated roses: faded petals, pruned stems. Wild roses: faded or fresh blooms, new shoots, fresh leaves picked just as they mature (June); hips (see special notes)
Processing: the petals may be separated by colour if you have enough. Otherwise, mix shades or use faded rose blooms with other flowers, such as petunias. Blooms can be fermented in a closed container filled with water and a pinch of baking soda. In this way you can add more petals as

they become available. The mixture can sit for several weeks before processing the dyebath. Although I use wild rose shoots early in the season, Lesch recipes recommend using fall cuttings (p 22). Spring cuttings are softer and more flexible. They are, therefore, easier to cut.
Colours obtained: mixed shades of hybrid tea petals, fermented: yellow (alum); medium brown (chrome); gold (blue vitriol); bright yellow-orange (tin); grey (iron). Fresh shoots (canes) and leaves, wild rose: yellow-green (alum); gold (blue vitriol); greenish-bronze (chrome); medium brown (iron); bright yellow-orange (tin)
Fastness: excellent for shades from canes and leaves; good for shades from fresh petals; good to excellent for shades from fermented petals
How to identify: cultivated varieties, Sherk and Buckley; photographs in nursery catalogues. (Rose enthusiasts tell me that the selection at McConnell Nurseries in Port Credit, Ontario, is superb. They feature colours which include blue, grey, and black.) Wild varieties, Cunningham; Gibbons; MacLeod and MacDonald; Petrides; Stewart and Kronoff
Availability: look for wild roses in rural areas, growing along or near roadsides and near fields, pastures, and mixed woods. Nursery stock is widely available.
Special notes: Lesch obtains black from rose bush trimmings, using 4 oz (120 ml) of iron as mordant for 1 lb of wool fibre (453 g). (Lesch 75. Other mordants used are tartaric acid and Glauber's salts.) I tried this using the amounts as given but found the fibre sample too crimpy. Rose hips are the bright orange-red fruits of the shrub that appear in the fall after flowering. They vary in size and colour, depending upon the species. However, all are uniformly high in vitamin C. Three hips contain more of this necessary nutrient than an orange (MacLeod and MacDonald 58). I prefer not to use them for a dye. I have tested a few, and the shades obtained were: a warm tan with vinegar and a pinkish-tan with chrome. Rose hips behave much like mountain ash berries in the dyebath, which is to say they cook up like a mush, and the dyed fibre retains particles even after the yarn has been worn.

St John's Wort

perennial wildflower, weed *Hypericum perforatum*

Dyers who have difficulty identifying St John's wort might keep in mind that it flowers earlier than its look-alike, ragwort (*Senscio jacobaea*), and does not grow as tall, although both plants have a similar habitat. St John's wort also has a much smaller leaf. Another identifying feature is the small dots which can be seen along the edges of the yellow petals when the flowers are held up to the daylight. Apparently St John's wort contains a toxic substance which causes a unique reaction to white-haired livestock. Frankton and Mulligan write that such animals, when exposed to strong sunlight after eating this plant, suffer from irritation and weight loss (p 122). Species to look for other than *H. perforatum* include *H. boreale* (swamp St John's wort); *H. mutilum* (dwarf St John's wort), and *H. canadense* (Canadian St John's wort).

Parts used: flowering tops; whole plant
Processing: as for whole plants
Colours obtained: flowering tops alone give yellows and golds; the whole plant: yellow-green (alum); bronze (blue vitriol); gold (chrome); medium yellow-green (iron and tin). Davidson writes that the leaves are reported to give red but the *Hypericum* species used is not given (p 18). The Krochmals list red from the leaves and flowers, but they describe a *Hypericum* species which grows to six feet (1.82 m) in height (p 171). This might be *H. pyramidatum* (great St John's wort), but they do not specify. A Swedish writer, Astrid Swenson, explains that the red colour comes from a tiny gland on the stalk of the plant ('A Dyeing Project in Sweden' in *Natural Plant Dyeing* 40). This contains a fluid which, she says, will stain your hands red when the flowers are crushed. Although the documented red-yielding species is *H. perforatum*, I have been unable to confirm this even though I have used that species for dyeing. To add to the puzzle, Robertson shares my inability to obtain red from *H. perforatum*. She repeats the Thurston report that St John's wort was traditionally used for red, but writes that she has been unable to verify this with her own experiments (p 47).
Fastness: excellent
How to identify: other species may be either more branched or less leafy than *H. perforatum*. See also Frankton and Mulligan; Peterson; Roland and Smith (they show maps of the Nova Scotia range for several species, p 514).
Availability: although St John's wort is common throughout much of Canada and the United States, it is somewhat localized, appearing abundantly in one region and not at all in another. It grows mainly in cultivated areas, often along roadsides adjacent to fields and pastures. The bloom appears in this region in late June or early July and lasts until early September.

Seaweed

See dulse, page 138.

Shasta Daisy

perennial, garden flower *Chrysanthemum*, various spp.

Shasta daisies are one type of chrysanthemum. There are many others, including ox-eye daisies and cushion mums. A few are annuals, but most are late-blooming perennials available as both dwarf and tall species. They range in colour from the white of Shasta daisies to yellow, orange, amber, rust, blue, and mauve. Any colour of bloom may be used, and flowers from different varieties can be combined in a single bath. As with all fresh flower baths, keeping a low temperature in the dyepot ensures the lightest and most delicate shades. (See p 65.)

Parts used: blooms, any colour
Processing: as for fresh flowers; below a simmer for light colours; simmer for stronger shades
Colours obtained: soft yellow (alum, below a simmer); yellow-gold (chrome); bright yellow (tin)
Fastness: excellent
How to identify: Herwig and photographs in most seed catalogues
Availability: as seed from CK, DO, MC, ST, and V. As bedding plants from most local nurseries.
Special notes: potted chrysanthemums are often featured seasonally as gift houseplants. However, these have so few blooms that they are only suitable for dyeing a very small amount of fibre. Also, because such plants are grown expressly for the commercial trade during the holidays, it is probable that the soil in which they are grown and other circumstances of cultivation would result in their producing different shades than mums from your own garden.

Sorrel

weed, perennial *Rumex acetosella*

Sorrel is a member of the same genus as dock (*R. crispus*, p 136). *R. acetosella* is sheep sorrel, and *R. acetosa* is garden sorrel. Leechman uses the latter species as a source of grey-blue (p 53), and Thurston uses various species for greens (p 6). Sorrel is not easily confused with dock because the leaves of each are very distinctive. Sorrel leaves are arrow-shaped and the plant has tiny green flowers. The seeds turn reddish-brown in the fall. A common plant in rural areas, Frankton and Mulligan write that it is rare in Manitoba, Saskatchewan, and Alberta (p 36). Both

R. acetosella and *R. acetosa* are abundant in Nova Scotia (Roland and Smith 351).

Parts used: the whole plant
Processing: as for whole plants
Colours obtained: soft yellow-green (alum and blue vitriol); medium dull green (blue vitriol, saddened in iron); chartreuse (tin); tan (chrome)
Fastness: excellent
How to identify: Frankton and Mulligan; MacLeod and MacDonald, Martin; Peterson
Availability: See above. *R. acetosa* is available as nursery stock from SH. Sorrel grows about 20 inches in height (50 cm) and prefers poor soil. *R. acetosa* is a larger plant, and may reach 3 feet (.92 m) in height. The young leaves are said to be a tasty addition to salads (MacLeod and MacDonald 44).
Special Notes: Davenport lists the roots of *R. acetosella* as a source of red (p 119). I tested this but was unable to obtain anything other than a pinkish-tan with alum and chrome as mordants.

Spearmint

See mint, page 177.

Spinach

annual, vegetable *Spinacia oleracea*

Although I have used spinach as a dyestuff when my own garden crop went to seed, many dye references consider colours from this vegetable to be fugitive. Davidson suggests it is not a worthwile dyestuff as so much of it is required to make a yellow-green (p 18). Leechman's recipe is a case in point: it uses 4 lbs (1.8 k) of spinach to dye 1 lb (453 g) of fibre, and that is a lot by anyone's standards! Grae does not even list the plant as a dyestuff, so she too must think it is not fast. Perhaps I have been lucky, but my spinach colours are fairly fast to light and washing. The tan is quite fast.

Parts used: fresh leaves or seed stalks
Processing: as for leaves. Chop or shred the spinach first. For yellow-greens, cover the spinach with water and allow it to soak out overnight. The following day, add the wetted fibre to the pot and raise the heat until

a slow simmer is reached. Let the fibre and spinach cook together for half an hour. Then remove the bath from the stove and allow it to sit overnight until completely cool. The following day, remove the fibre from the bath and rinse it in warm water to which you have added ½ cup of white vinegar (120 ml). Rinse the fibre again in cool water.
Colours obtained: the above procedure gives yellow with alum; bright yellow-green with tin; yellow-green with blue vitriol and gold with chrome. A medium bath was used (equal weight of fibre and spinach). For tan, use a vinegar mordant; for grey, use an iron mordant.
Fastness: the yellow-greens have fair light fastness; the tan and grey are quite fast to light and washing
How to identify: see the seed catalogues. Spinach and Swiss chard are quite similar, but chard has a larger leaf that is lighter green in colour.
Availability: from CK, DO, MC, ST, and V as seed. Spinach is easy to grow but needs good soil. It is just as useful when the plants have gone to seed. The Krochmals suggest using canned spinach. To those who know and love the flavour of fresh spinach, that may well be the best use for the tinned variety!

Spruce

coniferous ornamental, lumber tree *Picea*

Picea rubens, or red spruce, is a valuable tree for lumber in the northeastern United States and eastern Canada. There are several other species which are common in the same habitat, including *P. glauca* (white spruce, cat spruce), and *P. mariana* (swamp or bog spruce). The needles, branch tips, and cones of spruce are processed for dyeing as they are for fir (see p 142) and hemlock (see p 151). As is the case with all cones they may impart a stickiness to dyed fibre. Leechman (p 25) and the Krochmals (p 88) both list a yellow-orange from spruce cones with an alum mordant.

Parts used: needles, branch tips, cones, bark
Processing: according to type
Colours obtained: needles, used fresh: yellow (alum); gold (tin); bronze (chrome). Branch tips: bright yellow (tan); green (iron). Cones: reddish-tan (alum and chrome); brown (iron). Bark: yellow-tan (alum); reddish-brown (chrome, strong bath); taupe (iron)
Fastness: excellent for all
How to identify: Hosie; Knobel; Petrides; Saunders. The nursery catalogues have photographs of ornamental species.

Availability: MC, SH, and local nurseries. Wild trees can be transplanted in spring. Look for discarded Christmas trees for branch tips and visit sawmills for large amounts of bark.
Special notes: The yellow from spruce, fir, and hemlock tips is too strong to suit everyone's taste, but it is exceptionally fast and useful when top-dyeing with indigo for greens.

Squash

vegetable *Cucurbita*

Squash vines give a dye, as do fermented squash and pumpkin shells (the outer, hard covering). The vines are treated like those of the cucumber (see p 131). The skins or peelings from squash and pumpkin are soaked in water to cover after the fibrous part of the vegetable has all been scraped off the shell. Use enough water to just cover the peelings. Add a pinch of baking soda to eliminate some of the ensuing odour, and allow the covered container to sit in a warm place for two weeks. Cook out the mixture and strain off the peelings before dyeing.

Parts used: see above
Processing: see above
Colours obtained: the vines give tan with vinegar and grey-green with iron. Fermented squash peelings: yellow (alum); yellow-orange (chrome); bright yellow (tin); yellow-grey (iron); rust (chrome and tin)
Fastness: good for shades from the vines; excellent for shades from peelings
How to identify: see various nursery catalogues. There are many varieties of both pumpkin and squash. I have used green and yellow hubbards and buttercup.
Availability: as seed from DO, MC, ST, and V; from local seed suppliers. Hallowe'en pumpkin shells are suitable as a dyestuff. The soot on the inside of the hollowed-out shell is actually a mordant giving shades of brown and grey.

Sumac

shrub or small tree, ornamental *Rhus typhina*

Although Adrosko (p 47) and Leechman (p 53) use the traditional spelling, 'sumach,' Hosie (p 260) uses the newer form, 'sumac.' This spelling is closer to the actual pronunciation, which is 'soo-mack.' The sumac has been widely used as a dyestuff for centuries, although the species varies depending upon the country. ('Family Dyeing in Colonial New England,' *Dye Plants and Dyeing* 77) Thurston mentions the ground-up leaves of *Rhus coriaria* (p 33), which grows in Europe, while Davidson lists *R. glabra* (smooth sumac, p 18). Sumac has been used not only as a dyestuff, but as a mordant, and in the tanning of leather. Robertson gives a recipe called the alum-tannin method for mordanting cotton (p 88). *R. typhina*, the 'staghorn sumac,' is well named. Its unusual silhouette is antler-like in appearance and the branches are covered with a grey, velvety-looking substance like that which covers the horns of deer, elk, and caribou. Wild stands of sumac are common in eastern Canada and New England. Indeed, entire stretches of turnpike in southern Maine and coastal New Hampshire are lined with sumacs. In the fall, when the leaves and red fruiting cones of the tree are a bright red, sumac clumps are a glorious sight. Dyers worried about mistaking sumac for its poisonous relative, *R. vernix*, should check with either Hosie or Petrides (Hosie 260, 262; Petrides 134, 156) for identifying characteristics of the latter. It has white berries or fruit, instead of red, and in Canada does not grow east of Ontario and Quebec. Petrides gives the United States range, which includes southwestern Maine, but claims *R. vernix* should not cause too much of a problem because of its somewhat inaccessible swampy habitat. Staghorn sumac also prefers damp locations, but its conspicuous red fruiting cones and abundant growth in this region make it conspicuous. It may occur singly, but more often grows in a clump, ranging in height from 5 to 18 feet (1.5–4.5 m).

Parts used: fresh leaves, twigs or new shoots, mature limbs or bark, mature red fruiting cones (they are often referred to as 'berries'), or any combination of leaves, shoots, bark, and cones. Petrides refers to ink having been made from the boiled leaves and cones (p 134).
Processing: to use sumac as a mordant for cotton – collect an amount of fresh sumac leaves or leaves and shoots equal in weight to the fibre being dyed. Cook out the torn-up sumac, at a boil, for an hour (or until the water in the dyepot is yellow or yellow-green in colour). Remove the pot from the heat and strain off the sumac. Add the alum, which has been dissolved in boiling water. (For cotton, use 8 Tbsp or 120 ml of alum for 1 lb or 453 g of fibre.) Stir well until the alum is completely dissolved in the dyepot. Then add the wetted cotton, and proceed with the dyeing. This alum-tannin mordant method for cotton will actually colour the fibre a very pleasant shade of yellow or yellow-green. This will subsequently affect the colour the cotton is dyed. For instance, alum-tannin mordanted cotton dyed in an onion skin bath will be a brilliant orange or

rusty orange, if the mordant bath is yellow-green. If the sumac bath is yellow, however, the resulting onion skin bath would be a brilliant yellow. Process leaves as for fresh leaves; shoots, twigs, and bark as for bark; the cones are soaked out in water to cover to several days. Stir the mixture frequently, making sure all the cones are thoroughly wet. The water will turn rusty red or brown within two days, at which time the mixture is ready to be used for a dyebath.
Colours obtained: sumac is as prolific a dyestuff as the onion and the rhododendron with respect to the range of colours it yields. Fresh leaves give yellow with alum; brilliant yellow with tin; bronze with blue vitriol; bright gold with chrome; and yellow-green or khaki with iron. The twigs, shoots, and bark give yellow if the temperature of the bath is kept to a slow simmer (190°F, 88–90°C) and the processing time is relatively short. With higher heat and a longer cooking time, the shades are gold, tan, brown, bronze, and olive-green. The cones give reddish-brown with chrome; warm tan with alum, and a fine dark red with brown overtones with chrome saddened in iron. Cones processed with vinegar alone give pinkish-reds. Davidson gives a recipe for black from sumac leaves, twigs, berries, and nails, which nevertheless sounds fascinating (p 18). The Krochmals' Indian recipe for black is even more lengthy (p 27). Worst lists a grey from sumac with mulberry bark and iron (p 41).
Fastness: excellent for all, but pink-reds from the cones may fade slightly. It is interesting to note that Furry and Viemont write that they consider colours from sumac leaves to be fugitive (p 8; recipe, p 30). They do give recipes for the berries (cones) and consider these fast.
How to identify: sumac is pictured on the cover of *Dye Plants and Dyeing* and in that book on page 77. See also Hosie; Knobel; Petrides; Sherk and Buckley. Ask locally, especially at parks.
Availability: as nursery stock from SH; also from some local nurseries. The wild staghorn is easily transplanted, and since sumac sends up new shoots from the parent tree each spring, it reproduces rapidly. The cone-like fruiting bodies which top each stem remain on the tree all winter unless eaten by wildlife and birds. This is the most recognizable characteristic of the tree.

Sunflower

annual, domestic and wildflower *Helianthus annuus*

Sunflowers are tall-growing annuals often planted by home gardeners as a source of seed for their own use or as a wild bird food. In the western provinces and states, acres of nodding sunflowers are a common sight. They are commercially grown for sunflower oil. Reaching six feet or more in height (1.82 m), sunflowers are easily grown by beginning gardeners, and those uneaten seeds tossed out for finches and grosbeaks may well germinate. It is the flowers which are most often used, but the stalk and leaves can also be made into a dye. Lesch offers recipes using the seeds to obtain green and grey-blue (pp 86, 126). Many references say the colour obtained from the flowers is not fast, but Davidson suggests an after-rinse in an alkaline solution (such as lye soap) will solve this problem (p 19). I have not tried it, as my sunflower colours seem fast.

Parts used: flowers; stalks and leaves; seeds (see above)
Processing: treat flowers accordingly, following the dyeing with a rinse in an alkaline solution as suggested above. This could be a mild baking soda or iron dip, although these would probably change the yellow to a yellow-green or chartreuse. If this is not desired, first rinse as usual, which is what I do. Chop stalks and leaves as finely as possible, using a cleaver.
Colours obtained: from fresh flowers: yellow (alum and tin); gold (chrome); yellow-green (alum, after rinse in baking soda); chartreuse (alum, after rinse in iron, no heat); stalks and leaves: yellow-green (alum and tin); grey (iron); bronze (blue vitriol); seeds: tan (chrome)
Fastness: good for all
How to identify: Gibbons; Martin; Peterson; see nursery and seed catalogues
Availability: as seed from CK, DO, MC, ST; also from department and hardware stores, as bird seed
Special notes: H. annuus is the common domestic sunflower. In some areas, this has escaped to the wild. There are numerous wild species of varying size and flower shape, some of which resemble black-eyed Susans and coneflowers (see p 106).

Sweet Fern

woody shrub *Comptonia peregrina*

Read about gale (*Myrica gale*), page 145.

Sweet fern often grows in wild blueberry fields, where its distinctive odour is familiar to even the most casual picker. A rough, woody shrub growing from 2 to 4 feet in height (.61–1.22 m), sweet fern often appears to be a dirty-looking, dusty, brownish-green. This colour belies its potential as a dyestuff. The leaves are unlike those of *Myrica gale* in that they are 'scalloped,' whereas the gale leaf is shaped like the end of an oar or paddle. The foliage of sweet fern is so arranged on branches coming out from the main stems that the overall appearance of the shrub is fern-like. Aside from blueberry fields and other areas where the soil is acid, sweet fern grows along backroads and in ditches bordering fields and pastures. This habitat is different from that of *Myrica gale*, which can tolerate more moist conditions and a less acid soil. Whereas sweet fern can be found growing at the edges of wooded areas, gale often grows amid mixed woods. Where I live, sweet fern and gale grow within one hundred feet of each other, fern in a ditch at the edge of the woods, and gale within the woods, near spruce, hemlock, and birch.

Processing: treat sweet fern the same as *Myrica gale*, using the same parts of the plant
Colours obtained: the shades are very similar to those obtained from *Myrica gale*. Leaves only (picked in July): bright yellow (alum); brilliant yellow (tin); strong gold (chrome); warm, rich brown (blue vitriol and iron). Some twigs mixed in with the leaves will give good browns and greens with iron. The twigs and stems alone give orange-browns with chrome and tin in a strong bath.
Fastness: excellent
How to identify: see references for *Myrica gale*. Also, MacLeod and MacDonald; Stewart and Kronoff
Availability: from nurseries, for use as a groundcover to plant on banks to prevent soil erosion. Sherk and Buckley write that sweet fern is difficult to transplant (p 59). Read above and see *Myrica gale* for habitat. Of the two shrubs, I find sweet fern far more common in eastern Canada.
Special notes: there are several recipes (Robertson 36; *Dye Plants and Dyeing* 69) for *Myrica gale* in dyeing books but none that I could find for *Comptonia peregrina*.

Swiss Chard

See chard, page 117.

Tamarack

See larch, page 162.

Tansy

wildflower, herb, weed *Tanacetum vulgare*

An annual, tansy is often confused with tansy ragwort, *Senecio jacobaea* (200). However, while the ragwort blooms have petals, tansy flowers are flat-topped and button-like in appearance. Tansy also has a strong odour. Traditionally made into a tea and used as an embalming agent (Erichsen-Brown, *Herbs in Ontario* 29) it is now more often enjoyed as a garden flower. Although it is common throughout much of Canada and especially the eastern United States, tansy is quite localized (for example, near Nictaux, Annapolis Co, NS; tansy is also reported to virtually cover Navy Island in the Bedford Basin of Halifax Harbour each August). It grows freely from seed collected from the mature flowering heads, but can become a nuisance if allowed to establish itself where you don't want it to grow.

Parts used: flowering heads; leaves; whole plant
Processing: according to type
Colours obtained: flowers give strong yellows, gold, and warm yellow-tans; leaves give yellow-green and green; the whole plant gives yellow-green, bronze, and olive-green. Keep the temperature of the dyebath below a simmer for light yellows, and use a strong dyebath and iron for greens. Robertson (p 44) writes that she finds tansy ragwort makes a stronger dye than tansy but I have found the opposite to be true.
Fastness: excellent
How to identify: Frankton and Mulligan; Roland and Smith; Peterson
Availability: ask locally for seed from people who have established tansy clumps. The seed is also available from seed houses specializing in herbs (see suppliers, 228).

Tea

herbal beverage *Thea sinensis*

Read about coffee, page 125.

Like coffee, tea must be used with mordants to make the dyes obtained from it fast to light and washing. The colours from tea depend upon the type of tea used and the strength of the bath. Black tea gives the best dye, in the sense that it is the most fast. Used tea bags or fresh bags can be used, as well as fresh or used tea leaves. Tea and coffee can be combined in a single bath, and tea combines well in the dyepot with such herbs as mint (p 177) and chamomile (p 116). My favourite tea, Earl Grey, yields quite pale colours, but results from Orange Pekoe are diverse and interesting.

Parts used: used or fresh tea leaves, bags, or leftover beverage (without milk added to it); the addition of lemon or sugar will not matter (see p 75)
Processing: collect tea (in any form) and save it in a covered, plastic container until there is enough for a dyebath. Add additional water if necessary to keep bags thoroughly wet. Add a small amount of white or cider vinegar to reduce the odour. When you have enough, cook out for one to two hours. Strain off the bags (or leaves). Then add the wetted fibre and proceed with the dyeing.
Colour obtained: generally speaking, a medium tea bath of Orange Pekoe bags gives beige with alum; light tan with vinegar; tans and medium browns with chrome; taupe with blue vitriol and grey or beige-grey with iron. Rose tans sometimes result using alum and chrome as mordants.
Fastness: variable. Fibres left to cool overnight after dyeing in a tea or coffee bath seem to be more fast than those rinsed immediately.
Availability: both tea and coffee illustrate our society's conspicuous waste. Used bags and grounds are available from a variety of sources. Most organizations which sponsor suppers and teas will be happy to save the hundreds of bags they use for you if you stop by early in the day with a suitable container labelled with your name and address.
Special notes: an after-rinse in vinegar may help tea colours to be more fast. Most dyers have their own method of dyeing with tea, so experiment and find the one that suits you best.

Tomato

fruit *Solanum lycopersicum*

Tomato vines are a source of dye, and if you have an overabundance of green tomatoes, these too can be used in a dyebath. Fresh, ripe tomatoes are usually only used if they are diseased, frost-bitten, or otherwise spoiled for human consumption.

Parts used: tomato vine (after harvesting the fruit)
Processing: process as for cucumber and squash vines. If using green tomatoes and the vine together, soak them in water to cover for two days and then cook out. Strain off the dyestuff through an old cloth or screen, otherwise the seeds will stick to the fibre in the dyebath.

Colours obtained: fresh vines, with vinegar, give yellow-tan to tan; a good soft yellow with alum; a bright yellow with tin. Vines and green fruit together give yellow-green with alum and blue vitriol; medium dark green with iron. Frost-bitten vines and mature fruit together give a good brown with chrome and tan.
Fastness: good to excellent; somewhat variable using just fresh vines
How to identify: ask home gardeners to let you have their vines after the tomatoes are ripe
Availability: as seed from CK, DO, MC, ST, and V. As bedding plants from most nurseries specializing in fruit. There is a wide variety of tomato species. Large farms which produce this as a cash crop will have tons of vines left over, unless they are organic growers who then use this as a composting material. Most home gardeners who have a dozen plants will have enough vines for a few dyebaths.
Special notes: Lesch obtains red-browns using fresh tomato vines with a tin mordant (p 116).

Tulip

flowering bulb *Tulipa*

Petals from fading tulips make an excellent dye. They can be used alone or combined with other blooms. Petals of various colours can be mixed together for a dyebath and stored in water to cover in a plastic container until you have enough to use. Tulips combine especially well with daffodils and narcissi. Pick the flowers after they have faded, and place the container in a warm spot for a few days or even several weeks. Stir the mixture occasionally. It will be strong-smelling, but the colours produced in this manner are lovely. Add baking soda to the soaking mixture to help eliminate some of the odour.

Parts used: faded flowers
Processing: as above
Colours obtained: yellow and white tulips mixed with daffodils: yellow (tin); gold (chrome); yellow-grey (iron); red and maroon tulips: reddish-tan (alum and vinegar); grey (iron); warm brown (blue vitriol and chrome)
Fastness: good
How to identify: CK, MC, and SH, offer special fall catalogues featuring flowering bulbs. These carry photographs of literally dozens of varieties of plain and fancy tulips in every colour imaginable, including blue and black.

Availability: as bulbs for fall planting, tulips are available from most seed houses and department and hardware stores. Although their initial cost may seem high, the tulip bed multiplies yearly and provides the dyer with many flowers that require little or no maintenance.

Vetch

wildflower, weed *Vicia cracca*

Cow or tufted vetch is a tendrilled vine with many small blue or purple flowers. It spreads out to a considerable length, tangling itself among wild rose bushes, daisies, clover, and buttercups. In some places vetch forms almost a mat. When you walk through a hayfield, this clings to your legs and makes walking difficult. Although it is valuable as fodder for cows, vetch is troublesome when it invades cultivated land.

Parts used: the whole plant
Processing: as for whole plants. Tear the vine up and soak it out in water to cover for one or two days before processing.
Colours obtained: soft yellow (alum); tan (chrome); bright yellow (tin); gold (blue vitriol); green (iron). A strong bath is recommended.
Fastness: excellent
How to identify: Cunningham; Frankton and Mulligan; Martin; Peterson
Availability: vetch is in bloom at the same time as daisies, buttercups, and early clover. Look for it in hayfields. I have some which grows at the edge of hayfields intertwined with wild rose bushes and which lasts each year until October. Once identified, vetch is easy to remember.

Violet

wildflower *Viola*, various spp.

There are many types of violet, including those that are white, yellow, blue, and purple. The pansy is a member of the same genus. Some violets are wild, others are domestic garden flowers. All are useful for dyeing, but there may be a scarcity of some species of wild violets and dyers are urged not to collect these. As noted by MacLeod and MacDonald (p 8), the leaves are high in vitamin C. With the poor eating habits some of us have, it might be worthwhile to eat the leaves rather than make a dye from the flowers. The blooms are so small, it is unwise to use them for dyeing a large amount of yarn. Instead, dip a few small hanks into the violet bath and then use these for embroidery or tapestry work. I have tried to transplant wild violets with limited success. Some nurseries offer wild species.

Parts used: leaves and flowers: pick only if very abundant in a certain area
Processing: as for flowers (use leaves and flowers together)
Colours obtained: (using a sample hank weighing 1 oz or 28.25 g): yellow (alum); bright yellow-green (tin); tan (chrome); gold (blue vitriol); grey-green (iron)
Fastness: good
How to identify: Cunningham; MacLeod and MacDonald; Peterson
Availability: look for violets in May and June growing in open fields, meadows, or mixed woods. When collecting flowers and leaves, take care NOT TO PULL OUT THE ROOT of the tiny plant. From CK and DO as seed, MC as nursery stock
Special notes: MacLeod and MacDonald give a variety of interesting uses for violets, both culinary and medicinal (p 8).

Walnut

---

deciduous ornamental tree *Juglans nigra*

See page 112.

The black walnut and the butternut are closely related. Hulls from the black walnut have long been acknowledged as one of the finest sources of brown available, providing a deep, rich colour that is variable depending upon the mordants used, but always fast and true. Although not indigenous to eastern Canada, black walnuts are often planted here as ornamentals. If you do plant this tree, however, take care to locate it in the proper setting (Hosie 136; Petrides 135).

Parts used: fresh leaves, nuts, green nut husks or hulls, bark
Processing: the green hulls, which are very sticky to the touch, are picked in the early fall. Cover them with water, and soak out for several days prior to dyeing. Cook them at a high heat for an hour, and then strain off the dye liquor. The nuts are processed the same as the bark. These too benefit from soaking out prior to the dyeing. The bark is much stronger as a dyestuff than the nut within the husk, but the husks are the strongest of all.
Colours obtained: leaves: yellow (alum); bright yellow (tin); gold (chrome); brown (chrome and iron). Hulls: brown (chrome); dark brown (chrome

and iron); taupe (blue vitriol); warm tan (alum); bright rusty brown (alum, bloomed in tin). The nuts give beige to tan. The bark gives brown with chrome, but it is not as rich as the shade from the hulls. See special notes.
Fastness: excellent
How to identify: Hosie; Knobel; Petrides. Ask locally, especially at parks.
Availability: MC offers Carpathian walnut (*Juglans regia*), which is a strain of English or Persian walnut. SH offers this too, and, in addition, *J. nigra*. Their catalogues make no mention of the toxic root substance.
Special notes: Davidson gives a recipe for obtaining black from fall walnut leaves (p 20), and Thurston suggests first dyeing with indigo and topping with walnut for a black from the hulls (p 26). It is interesting, and perhaps sad as well, that people who own black walnut trees suffer damage to them when would-be dyers rob these of their hulls each fall. People who have walnuts tell me they do not mind allowing others to pick the hulls, if they ask, and do it in the daylight. But it seems as if some people who use the hulls for dyes prefer to do their collecting in the dead of night. In the process, they often damage the branches and surrounding gardens. Most gardeners are fair-minded and generous. Taking the liberty of ruining their property only causes alienation and hard feelings. Most owners of walnut trees will offer the hulls to you if you can demonstrate reasonable courtesy when picking them.

Willow

deciduous tree, wild and ornamental *Salix*, various spp.

There are many varieties of willow, some of which are difficult for the amateur to identify. Roland and Smith (p 321) write that willows are highly variable and tend to hybridize. The common and easily recognized pussywillow is *S. discolor* or *S. humilis*, neither of which reaches tree size. The ornamental weeping willow is *S. alba*. Most dye references refer to *S. negra*, black willow, as its bark is said to yield a distinctive rose-tan with an alum mordant (Furry and Viemont 12). The addition of iron and blue vitriol will produce black, according to Davidson, but she thinks that colours from willow leaves are not fast (p 20). I have tried the leaves from weeping willow, however, and found them to be fairly fast. The only thing to do, then, is to try various willows and decide for yourself.

Parts used: leaves (see above); twigs and bark
Processing: according to type
Colours obtained: from the leaves of weeping willow: yellow (alum,

strong bath); yellow-green (alum); bright yellow (tin); greenish-yellow (blue vitriol); gold (chrome); soft light grey (iron)
Fastness: good for leaves and excellent for twigs and bark
How to identify: Hosie (Hosie refers to weeping willow as *S. babylonica*, but Petrides and the Sheridan catalogue list the weeping species as *S. alba* 'Triste'); Knobel; Saunders; Petrides. Ask locally. See below.
Availability: MC and SH as nursery stock; available at most local nurseries. Willows are often more easily identified by the novice in the winter or spring. The barks of most are decidedly yellow or reddish in colour. Pussywillows are apparent from February on, depending upon the locale. Most smaller willow species grow in clumps, and prefer fields, meadows, pastures, and the edges of mixed woods. Willows grow amid alder, chokecherry, elderberry, and mountain ash. Saunders (p 42) mentions that willow and poplars are said to be able to take root successfully when a fresh cutting is stuck into moist ground. However, I have had variable results with this method of propagating these species.

Woad

biennial *Isatis tinctoria*

Although not native to this country, woad is still found in the wild in England. The blue colour obtained from it is not always fast, nor is it as strong in hue as that from indigo. However, efforts on the part of Nova Scotian spinners, weavers, and dyers to cultivate this fascinating traditional dyeplant deserve to be mentioned here. Although Robertson (p 79) gives planting instructions for woad, hers apply to Britain. I offer instead the advice given to me by Carol Duffus, a weaver who lives in Bedford, NS, and Elfriede Budgey, a Truro spinner.

Directions for planting woad: wait until the weather has stabilized and the soil is warm. (For Mrs Budgey, this is after mid-May.) Choose a location that is sunny, well cultivated, and well drained. Woad prefers a sandy soil that is slightly alkaline rather than acid. Make rows about ½ inch (1.3 cm) deep and sow the seed. Keep the area damp until the seeds germinate and the small plants are well established. The first year, the woad plant will produce only a flat rosette of light green, shiny leaves. This dies down in the winter and the following spring the flower stalk appears. It will reach a height of 4½ feet (1.35 m) in this area, and have small clusters of yellow flowers at the top. The blooms last 3 to 4 weeks, and are followed by purplish seedpods. Collect these seeds when mature for replanting the following year.

Parts used: fresh leaves are used to make the dye, and are considered best if picked just as they reach maturity, in late July or early August.
Processing: woad must be fermented and prepared in a vat dye as indigo (see p 154) in order to be fast to light and washing. However, a non-fast method nevertheless interesting as an experiment has been tried by Alicia Marr, who has also experimented extensively with lichens. She adapted this method from those she read about in several books. She covers the torn-up leaves with boiling water in a screw-top jar. This is allowed to sit for half an hour. Then she adds a small amount (1–2 tsp, 5–10 ml) of ammonia and shakes the mixture. Remove the jar lid and dip in the wet fibre (obviously a small amount if using a jar). You can use more fibre if you make up the mixture in a larger vessel. The mixture and the fibre are green until you lift the yarn out to oxidize it, whereupon it turns blue. Successive dips may be made to deepen the colour. Mrs Marr obtained a lovely soft 'sky blue' shade that was not fast to light and washing. Nevertheless, such yarn might be useful as a base for a subsequent colour.
Processing of the woad vat: because my woad plants are still immature, I have been unable to test the woad vat method myself. The few scattered instructions I have read seem complicated. Robertson (p 79) gives the information from William Rhind's 1857 *History of the Vegetable Kingdom*, and suffice it to say that only the most dedicated dyer would pursue this approach. Robertson's recipe for the non-vat method is not unlike Mrs Marr's except that Robertson adds bran and lime to the mixture in the jar and keeps the temperature close to 100°F (38°C) for 6 to 12 hours.
Colours obtained: the blues from woad require no mordants. They are said to be softer and less strong than those from indigo.
Availability: Mrs Duffus's seed originally came from a friend in England, but now that her own woad patch is established, she has been generous in distributing what seed she produces. See suppliers, page 228. Mrs Budgey ties paper bags over the flowers as they are turning to seed to protect them from flying away.

Yarrow

perennial, wild and cultivated flower *Achillea*, various spp.

Yarrow (*A. millefolium*) is often mistaken for Queen Anne's lace (*Daucus carrota*) which it superficially resembles, as both have white, flat-topped flowering heads and fern-like leaves. However, yarrow blooms much later in the season and it has a white stem. Yarrow smells good, and, dried, is often used indoors for winter flower arrangements. Another wild species, sneezeweed yarrow (*A. ptarmica*), has more daisy-like flowers and narrow, saw-toothed leaves. Cultivated yarrows are usually yellow although one species, *A. argentea*, is white.

Parts used: flowering heads or the whole plant
Processing: according to type
Colours obtained: from *A. millefolium*, whole plant: bright yellow (alum); brilliant yellow (tin); strong, clear gold (chrome); yellow-tan (blue vitriol); yellow-grey (iron). Grae gives a recipe for yarrow using a tin mordant and processing the bath in an iron pot (p 111). Yarrow is a strong dyeplant and, even using a weak bath, will give excellent yellows and golds.
Fastness: excellent
How to identify: Cunningham; Frankton and Mulligan; Martin; Peterson
Availability: a variety of species are available from SH as nursery stock. Look for yarrow from July through late September, and possibly earlier in most of New England. Although a very common wildflower, yarrow is somewhat localized. I find it growing in clumps in ditches near fields and woodlots. It prefers the company of another late-blooming wildflower, goldenrod.

Zinnia

annual, garden flower *Zinnia*

Zinnias are popular garden annuals that vary in size and colour depending upon the species. Some are dwarf types, while others are tall-growing. Colours available include yellow, gold, maroon, crimson, magenta, mauve, and purple. As is the case with almost all popular flowers, hybridization is producing new strains, which results in more shades being available. Like marigolds and petunias, zinnias bloom through the entire season if you remove the faded flower heads.

Parts used: flowers, separated by colour or used mixed
Processing: as for flowers. Zinnias may be soaked for an extended period to produce darker, richer shades.
Colours obtained: in a strong bath of mixed colours: yellow (alum); bronze (chrome); bright gold (tin); khaki (blue vitriol); grey-green (iron)
Fastness: excellent
How to identify: see the photographs in CK, DO, MC, ST, and V catalogues. Ask locally.
Availability: as seed from the above, and as bedding plants from nurseries

Appendix

Energy-saving procedures

At a time when all types of fuel and other energy sources are increasing in price, it is important for the dyer to practise conservation. While most of these points have been mentioned before, read the list and make a mental note to be less wasteful next time you process a dyebath.

- Dye as much fibre as you can during each dyeing session.
- Avoid dyeing at periods of high energy consumption: work in the mid-afternoon or evening if possible.
- Soak out all dyestuffs prior to processing as this will lessen the time taken to extract the pigment. This is especially so when dealing with tough dyestuffs such as roots, nuts, and barks.
- Energy is wasted if a large electric burner is used to heat a very small pot. Try to select a pot which is as large as, or larger than, the burner. Keep the pot covered as much as possible.
- Hot plates are expensive to operate. If you must use one, get the single-burner type, as most pots are too small to fit over both burners at once on a two-burner unit. Where propane heat is available and less expensive than electricity, dyers should investigate using propane camp stoves.
- While dyeing and cooking should not be carried out at the same time, in the same space, it is possible with some organization to arrange to put your dyepot on the heat right after you take off the soup kettle. Dyers who have wood stoves can set the dyepot at the back and simply move it forward when they are through cooking. However, do this only if you are cooking out a dyestuff, and not actually doing the dyeing. The use of a wood stove also enables the dyer to put a dyestuff on to cook out just after the evening meal has been prepared. The pot can sit on the stove the entire evening, benefitting from what residual heat remains, even if that stove is not stoked up again.
- Rather than boiling extra water in which to dissolve mordants, take hot water from the pot in which the dyestuff is cooking out and

use that instead. Once dissolved, the mordant solution can be set aside in a safe place provided you are ready to use it within, say, an hour.
- Have the fibre ready to be dyed soaking out in very hot water. Then when you put the fibre in the pot the dyebath will not take as long to heat and you can cut down on the processing time.
- If someone in the family is having a bath, or you have fairly clean, warm dishwater left over in the sink, soak out the fibre to be dyed in that. The water should be free of, say, bath oils, or, in the case of dish-water, greasy food clumps.
- After removing the fibre initially dyed, put in additional skeins to take advantage of what exhaust pigment remains. Even if these skeins dye a very light shade, they can be topped later with another colour.
- To save on hot water for rinsing, leave the dyed fibre in the bath over-night and rinse the following day when it is cool.
- Drying dyed fibre inside the home humidifies it much like taking a shower will do. If you are dyeing a large amount of fibre in the winter and have a humidifier, unplug it for a couple of days while the yarn dries.
- Even in winter, a pot full of cold water can be warmed up considerably by placing it on a table in front of a sunny window. (The many possi-bilities of solar dyeing are explored in an article written by Marilyn Lorance for *Shuttle, Spindle & Dyepot*. See bibliography. Also see p 172.) Some dyers say that filling pots with hot tap water cuts down the time it takes for the pot to heat and they think that this represents an energy-saving technique. I cannot say that it does or does not. Perhaps the thing to look at is how the hot water you use is heated. If your hot water is cheap, fine; if it is expensive, find a way to heat cold water using the sun.
- If you have an outdoor barbeque or grill, put the dyepot on it after you are through cooking. If the grill is dirty, place an oven rack or piece of foil between it and the dyepot. A dyepot can be placed in a bed of coals in an outdoor fire so long as possible discolouring of the outside of the pot does not bother you.

Becoming a gardener

Dyers who love their craft inevitably become gardeners. Some actually start out as avid gardeners, and dyeing is taken up as another aspect of that initial interest in plants. But most of us grow some flowers, vege-tables, and a few trees without consciously planting specific species which will yield a dye. However, we can easily shift the focus of our gardening and plant species which enhance our living space and our fibre work.

The easiest plants for the novice gardener to deal with are annuals. They are purchased in April, May, or June (depending upon the region you live in) in jiffy pots, which are fibrous containers usually made of pressed peat moss. City dyers can place the flowers in window boxes or even in standard plastic or clay flower pots if they do not have a garden. Most annuals such as petunias, pansies, and marigolds require lots of sun and minimum care. If you are worried about transplanting them into flower pots, ask the nursery staff if they recommend leaving the plants in their jiffy containers. Some people who live in apartments grow annuals successfully in this way.

Perennials are useful flowers for the gardener who likes to know that the oriental poppies, coreopsis, lupin, and day lilies will come up every year without any attention other than perhaps a light winter mulch of straw or evergreen boughs. Ask the nursery staff to recommend perennials which are easy to grow. Each year new types appear on the market, and just about every kind of garden flower will give a colour. Perennials are more expensive, per plant, than annuals, but they are a good investment.

Trees, ornamental shrubs, and other types of foliage plants that give dyes are not necessarily hard to grow but are costly. (Transplanting from the wild requires you to know who owns the land; obtain their permission first.) If you have the space, first try those plants which will enhance the property as well as provide a dyestuff, such as privet. Even a single poplar will provide enough leaves for dyeing each spring, but you will require several large rhododendrons in order to harvest the leaves annually and not harm the shrubs. The cost of such ornamentals is high, but most reputable nurseries guarantee their stock and replace plants that fail to thrive, provided you have followed their planting instructions.

Read books on gardening and talk to other gardeners. Dyers are wonderfully keen on sharing plants, seeds, and hints about growing dyeplants. Take advantage of their generosity. Joining gardening clubs is an excellent way for the novice gardener to benefit from other people's gardening experiences without spending a lot of money and perhaps learning the hard way. For instance, if you have soil with a pH of more than 7 (alkaline), it is unlikely that acid-loving plants such as the rhododendron will thrive. Don't fight nature: move with it, and plant things that will grow in your soil or modify the pH (for more acid, add peat moss; more alkaline, add lime).

Experienced and novice gardeners as well make their task easier by having their garden soil tested regularly to determine the pH level. For information about soil-testing ask at the local nursery. Experienced gardeners often test their own, using commercial kits, but the local branch of the Department of Agriculture can advise you where to have this done, often at no cost.

Make friends of gardeners in your neighbourhood. It is amazing how interested gardeners become in your dyeing, to the point where they will deliver endless plant bounty to your doorstep. They enjoy knowing that you appreciate having it and know how to use it to produce something unique and beautiful. Gardeners and dyers are similar kinds of people. The best of both types love plants and do all that they can to keep this earth green and lovely for the future.

Metric conversion tables

WEIGHT OF FIBRE AND DYESTUFFS

1 ounce equals 28.349 grams, but to facilitate measurements, metric equivalents are given to the nearest tenth.

.035 oz	= 1 g	8 oz	= 228 g
1 oz	= 28.5 g	1 lb	= 453 g
4 oz	= 114 g	2.2 lb	= 1 kilo

MEASUREMENTS FOR MORDANTS AND CHEMICALS

¼ tsp	= 1 ml	⅓ cup	= 70 ml
½ tsp	= 2 ml	½ cup	= 120 ml
1 tsp	= 5 ml	1 cup	= 240 ml
2 tsp	= 10 ml	1 pt	= .5 l
1 tbsp	= 15 ml	1 qt	= 1.1 l
¼ cup	= 60 ml	35.2 fluid oz	= 1 l

VOLUME OF WATER FOR DYEBATH
(see p 63)

1 gallon	= 4.5 l	6 gallons	= 27 l
2 gallons	= 9 l	10 gallons	= 45.5 l
4 gallons	= 18 l		

SIZE OF DYEPOTS

See volume of water for dyebath. Most standard enamel 'canning' pots hold 4 gallons, or 18 l of fluid. 'Stock pots' tend to be larger, and are available in sizes up to 10 gallons, or 45.5 l. Copper boilers hold that much if not more.

TABLE FOR LENGTH

Formula for converting feet to metres: feet $\times$ 12 = $\frac{\text{inches}}{39.34}$ = metres

For example: 10 feet $\times$ 12 = 120 inches, divided by 39.34 = 3.0503 or 3 metres

1 inch	= 2.54 cm	4 feet	= 1.2 m
1 foot	= .3 m	5 feet	= 1.5 m
1 yard	= .9 m	10 feet	= 3 m
2 feet	= .6 m	50 feet	= 15.2 m
3.2 feet	= 1 m	100 feet	= 30.5 m

MORDANT AND DYEBATH TEMPERATURES

Do not use soap at warmer than 120°F or 50°C.
A slow simmer is 190°F or 88–90°C.
A simmer is 200°F or 91–95°C.
A boil is 212°F or 98–100°C.

Suppliers

KEY

F: fibres for dyeing; D: dyes and mordants; E: spinning and weaving equipment; B: books on dyeing

Albion Hills School of Weaving, Spinning & Dyeing
Caledon East, Ontario
L0N 1E0 F, D, E

Anja's Weaving Supplies
PO Box 10
RR 1, Chelsea, Quebec
J0X 1N0 F, E, B

Briggs and Little Woolen Mills
York Mills
Harvey Station, New Brunswick
E0H 1H0 F

Chamomile Shop
PO Box 619
Rangeley, Maine
04970 F, D, E, B

Condons Yarns
PO Box 129
Charlottetown, Prince Edward Island
C1A 7K3 F

Frederick J. Fawcett Inc.
129 South Street
Boston Massachusetts
02111 F, E

Gina Brown Needlecraft Studios
2207 4th Street SW
Calgary, Alberta
T2S 1X1 F, E

Greenmont Yarns & Looms
Bennington, Vermont
05201 F, E, B

Handcraft Wools
PO Box 378
Streetsville, Ontario
L5M 2B9 F, D, E, B

Harrisville Designs Inc.
PO Box 51
Harrisville, New Hampshire
03450 F, E

Island Crafts
335 George Street
Sydney, Nova Scotia
B1P 1J7 F

The Mannings Handweaving School
RD 2
East Berlin, Pennsylvania
17316 F, D, E, B

New England Earth Crafts
149 Putnam Street
Cambridge, Massachusetts
02239 F, D, E, B

Newfoundland Weavery
170 Duckworth Street
St John's, Newfoundland F, E
A1C 1G2

Nilus LeClerc Looms
CP 69
L'Islet, Quebec
G0R 2C0 E

Romni Wools
319 King Street West
Toronto, Ontario
M5V 1J5 F, D, E, B

Romni Wools
3779 W 10th Avenue
Vancouver, British Columbia
V6R 2G5 F, D, E, B

Valley Fibres
51 William Street
Ottawa, Ontario
K1N 6Z9 F, E, B

Vavstolsfabriken Looms
28064 Glimakra
Sweden
12979 E

The Village Weaver
551 Church Street
Toronto, Ontario
M4Y 2E2 F, E, B

Waltoncraft Studio
58 Dutch Village Road
Halifax, Nova Scotia
B3L 4E6 F, E, B

The Weavers Place
109 Osborne Street
Winnipeg, Manitoba
R3L 1Y4 F, D, E, B

Wide World of Herbs
11 St Catherine Street East
Montreal, Quebec
H2K 1K3 D

Wide World of Herbs
PO Box 266
Rouses Point, New York D
10034

SEED AND PLANT SUPPLIERS

See page 88.

C.A. Cruickshank Limited
1015 Mount Pleasant Road
Toronto, Ontario
M4P 2M1 wildflowers; seeds

The Sower
PO Box 159
Bear River, Nova Scotia
B0S 1B0 dyeplant seeds

Thompson and Morgan Inc.
PO Box 24
Somerdale, New Jersey
08083 dyeplant seeds

Bibliography

BOOKS

Adrosko, Rita J. *Natural Dyes and Home Dyeing* New York: Dover Publications, Inc. 1971

Bearfoot, Will *Mother Nature's Dyes and Fibres* New York: Charles Scribner's Sons 1975

Beriau, O.A. *Home Weaving* Gardenvale, Que.: Institute of Industrial Arts 1939

Bigelow, Howard E. *Mushroom Pocket Field Guide* New York: Macmillan Publishing Co. Inc. 1974

Black, Mary E. *New Key to Weaving* Toronto: Collier-Macmillan Canada Ltd 1972

– and Bessie Murray *You Can Weave* Toronto: McClelland and Stewart 1974

Bolton, Eileen *Lichens for Vegetable Dyeing* McMinnville, Ore.: Robin and Russ Handweavers 1972

Burnham, Dorothy and Harold *'Keep Me Warm One Night': Early Handweaving in Eastern Canada* Toronto: University of Toronto Press 1972

Canada Weed Committee, editors *Common and Botanical Names of Weeds in Canada* Publication 1397. Ottawa: Canada Department of Agriculture 1975

Crockett, James Underwood and Oliver E. Allen and editors, Time-Life Books, *Wildflower Gardening* Alexandria, VA: Time-Life Books, Time-Life Encyclopedia of Gardening 1977

Chipman, E.W. *Growing Savory Herbs* Publication 1158. Ottawa: Canada Department of Agriculture 1968

Culpeper, Nicholas *Culpeper's Complete Herbal* London: W. Foulsham & Co. Ltd 1972

Cunningham, G.C. *Forest Flora of Canada* Bulletin 121. Ottawa: Department of Northern Affairs and Natural Resources, Forestry Branch 1975

Cunningham, Robert and John B. Prince *Tamped Clay and Saltmarsh Hay (Artifacts of New Brunswick)* Fredericton, NB: University Press of New Brunswick, Ltd 1976

Davenport, Elsie G. *Your Handspinning* Pacific Grove, CA: Craft and Hobby Book Service 1971
– *Your Yarn Dyeing* Pacific Grove, CA: Craft and Hobby Book Service 1972
Davidson, Mary Frances *The Dye Pot* Gatlinburg, Tenn.: Mary F. Davidson 1970
Donkin, R.A. *Spanish Red: An Ethnogeographical Study of Cochineal and the Opuntia Cactus* LXVII. Part 5. Transactions of the American Philosophical Society, September 1977. Philadelphia: American Philosophical Society 1977
Eighty Years' Progress, 1781–1861 in the United States and Canada New York: New National Publishing House 1864
Erichsen-Brown, Charlotte *Herbs in Ontario* Aurora, Ont.: Breezy Creeks Press 1975
Erskine, J.S. *Common Lichens* Reprint from *The Journal of Education* April 1957. Halifax: Nova Scotia Museum 1958
Foster, Malcolm Cecil *Annapolis Valley Saga* Windsor, NS: Lancelot Press 1976
Frankton, Clarence and Gerald A. Mulligan *Weeds of Canada* Publication 948. Ottawa: Canada Department of Agriculture 1974
Furry, Margaret S. and Bess M. Viemont *Home Dyeing with Natural Dyes* Santa Rosa, CA: Thresh Publications 1975
Gerber, Frederick H. *Indigo and the Antiquity of Dyeing* Ormond Beach, FL: Frederick H. Gerber 1977
Gibbons, Euell *Stalking the Wild Asparagus* Field Guide Edition. New York: David McKay Co., Inc. 1973
Grae, Ida *Nature's Colors: Dyes from Plants* New York: Macmillan Publishing Co. Inc. 1974
Green, Gordon, editor *A Heritage of Canadian Handicrafts* Toronto: McClelland and Stewart 1967
Groves, J. Walton *Edible and Poisonous Mushrooms of Canada* Publication 1112. Ottawa: Research Branch, Canada Department of Agriculture 1972
– *Mushroom Collecting for Beginners* Publication 861. Ottawa: Canada Department of Agriculture 1973
Hale, Mason E. *How to Know the Lichens* Dubuque, IA: Wm Brown Co. Publishers 1969
Herwig, Robert *128 Garden Plants You Can Grow* New York: Collier Books, Macmillan Publishing Co. Inc. 1976
Hosie, R.C. *Native Trees of Canada* 7th ed. Ottawa: Canadian Forestry Service, Department of the Environment 1973
Journal of the Chicago Horticultural Society 111, Winter 1976. Chicago: Chicago Horticultural Society 1976
Kierstead, Sallie Pease *Natural Dyes* Boston: Bruce Humphries, Inc. 1950
Knobel, Edward *Identify Trees and Shrubs by Their Leaves* New York: Dover Publications Inc. 1972
Kramer, Jack *Natural Dyes: Plants and Processes* New York: Charles Scribner's Sons 1972

Krochmal, Arnold and Connie Krochmal *The Complete Illustrated Book of Dyes from Natural Sources* Garden City, NY: Doubleday & Co. Inc. 1974

Langdon, Eustella *Pioneer Gardens at Black Creek Pioneer Village* Toronto: Holt, Rinehart and Winston of Canada Ltd 1972

Lathrop-Smit, Hermine *Natural Dyes* Toronto: James Lorimer & Co. 1978

Leechman, Douglas *Native Tribes of Canada* Toronto: W.J. Gage Ltd, nd

– *Vegetable Dyes from North American Plants* Toronto: Southern Ontario Unit of Herb Society of America 1969

Lesch, Alma *Vegetable Dyeing* New York: Watson-Guptill Publications 1974

MacLeod, Heather and Barbara MacDonald *Edible Wild Plants of Nova Scotia* Halifax: Nova Scotia Museum 1976

MacPhail, Margaret *The Bride of Loch Bras d'Or* Windsor, NS: Lancelot Press 1974

Martin, Alexander *Weeds* Golden Nature series. New York: Golden Press 1972

McGrath, Judy Waldner *Dyes from Lichens and Plants* Toronto: Van Nostrand Reinhold Ltd 1977

Merrill, G.K. *Report of the Canadian Arctic Expedition, 1913–1918* IV, Part D: Lichens. Ottawa: National Museums of Canada 601-254. King's Printer 1924

Mitchell, Lillias *The Wonderful World of the Weavers* Dublin, Ire.: Lillias Mitchell and Department of Education 1972

Nelson, Peter K. *The Hundred Finest Trees and Shrubs for Temperate Climates: A Handbook* Plants and Garden Series 13, 3. Brooklyn, NY: Brooklyn Botanic Gardens 1973

Peterson, Roger Tory *A Field Guide to Wildflowers* Peterson Field Guide series. Boston: Houghton Mifflin Co. 1968

Petrides, George A. *Field Guide to Trees and Shrubs* 2nd ed. Peterson Field Guide series. Boston: Houghton Mifflin Co. 1976

Porter, Eliot *Summer Island* Abridged. Sierra-Ballantine Book series. New York: Ballantine Books 1968

Rice, Miriam C. *Let's Try Mushrooms for Color* Santa Rosa, CA: Thresh Publications 1974

Robertson, Seonaid *Dyes from Plants* New York: Van Nostrand Reinhold Ltd 1973

Roland, A.E. and E.C. Smith *The Flora of Nova Scotia* Halifax: Nova Scotia Museum 1969

Saunders, Gary *Trees of Nova Scotia* Bulletin 37. Halifax: Nova Scotia Department of Lands and Forests 1970

Sauvé, Paulette-Marie *La Teinture naturelle au Québec* Montréal: Les Editions d'Aurore 1977

Schetky, Etheljane McD., editor *Dye Plants & Dyeing: A Handbook. Plants & Gardens* XX, 3. Brooklyn, NY: Brooklyn Botanic Gardens 1973

Scott, Peter J. *Some Edible Fruits and Herbs of Newfoundland* St John's: Breakwater Books 1978

Sherk, Lawrence C. and Arthur R. Buckley *Ornamental Shrubs for Canada* Publication 1268. Ottawa: Research Branch, Canada Department of Agriculture 1968

Shuttleworth, Floyd S. and Herbert S. Zim *Non-Flowering Plants* Golden Nature series. New York: Golden Press 1976

Sparling, Mary *A Guide to Some Domestic Pioneer Skills* Halifax: Nova Scotia Museum 1972

Stewart, Anne Marie and Leon Kronoff *Eating from the Wild* New York: Ballantine Books 1975

Teuscher, Henry, editor *Handbook on Conifers. Plants & Gardens* XV, 2. Brooklyn, NY: Brooklyn Botanic Gardens 1969

Thresh, Robert and Christine Thresh *An Introduction to Natural Dyeing* Santa Rosa, CA: Thresh Publications 1972

Thurston, Violetta *The Use of Vegetable Dyes* Bristol, Eng.: Dryad Press 1970

Van Dam, Tony 'The Maritime Gardener' CBC radio transcript 79-3-11. CBC Radio Maritimes, Halifax, winter 1979

Weigle, Palmy, editor *Natural Plant Dyeing: A Handbook. Plants & Gardens* XXVI, 2. Brooklyn, NY: Brooklyn Botanic Gardens 1973

Wells, Oliver N. *Modern and Primitive Salish Weaving* Sardis, BC: Oliver N. Wells 1969

Wigginton, Eliot, editor *Foxfire 2* New York: Anchor Books, Anchor Press/Doubleday 1973

– *Foxfire 3* New York: Anchor Books, Anchor Press/Doubleday 1975

Worst, Edward F. *Dyes and Dyeing* reprinted in *Foot-Power Loom Weaving* Pacific Grove, CA: Craft & Hobby Book Service 1970

ARTICLES

Aiken, Marie 'Lichens as a Dye Source' *Craftsman/L'Artisan* III, 3 (1970)

Belson, Florence 'Beauty in Ontario's Pioneer Coverlets' *Canadian Collector* II, 1 (January 1967)

Blackburn, Edna 'Colours from Nature' *Canadian Collector* XIII, 3 (May/June 1978)

Burnham, Harold B. 'Niagara Coverlets' *Canadian Collector* I, 3 (Summer 1966)

Casselman, Karen 'Flowers into Gold' *Bluenose Magazine* I, 2 (Fall 1976)

– 'The Primeval Dyepot' *Harrowsmith*, 21 (Summer 1979)

– 'Winter Dyeing with Umbilicate Lichens' *Shuttle, Spindle & Dyepot* IX, 2 (Spring 1978)

Eggleston, Phyllis 'The Chilkat Blanket' *Shuttle, Spindle & Dyepot* VI, 1 (Winter 1975)

Gerber, Frederick 'Investigative Method: A Tool for Study' *Shuttle, Spindle & Dyepot* VI, 2 (Spring 1975)

Held, Shirley 'Exotic Woods for the Dyepot' *Shuttle, Spindle & Dyepot* IX, 4 (Fall 1978)

Koehler, Glory Dail 'Dyepot: Carcinogenic Properties of Fibre Dyes' *Shuttle, Spindle & Dyepot* VII, 4 (Fall 1976)

Lamb, I. MacKenzie 'Lichens of Cape Breton Island' *Annual Report of the National Museum of Canada for 1952–3* Bulletin 132, National Museum of Canada. Ottawa: Department of Northern Affairs and National Resources 1954

Lohmolder, Jo 'Dyeing with Bark' *Shuttle, Spindle & Dyepot* V, 2 (1974)

Lorance, Marilyn 'Nature's Colors, Naturally (Solar Dyeing)' *Shuttle, Spindle & Dyepot* X, 1 (Winter 1979)

McGrath, Judy Waldner 'The Dye Workshop' *North Magazine* XXII, 2 (March/April 1974)

MacNutt, A. Dawn 'Fibre Landscapes' *Canada Crafts* IV, 5 (June/July 1979)

– 'A Thing about Trees' *Shuttle, Spindle & Dyepot* IX, 4 (Fall 1978)

Ottawa Valley Weavers Guild 'Dyeing with Cochineal' *Ontario Spinners and Weavers Bulletin* XXI, 2 (Winter 1977)

Ponting, K.C. 'Kermes and Cochineal' *The Weavers Journal* Federation of British Craft Societies 108 (Winter 1978)

Pulleyn, Robert, editor *Fiberarts Dyes & Dyeing: A Special Issue* 1 (1978)

Shaw, Lynn 'Lichens as Dyestuffs' *Shuttle, Spindle & Dyepot* V, 3 (Summer 1974)

Taggart, Barbara 'Control Dyeing with Indigo' *Shuttle, Spindle & Dyepot* V, 3 (Summer 1974)

Walbridge, T. 'Dyepot: The Dye Garden' *Shuttle, Spindle & Dyepot* VIII, 2 (Spring 1977)

Indexes

Common names

Botanical names

Colour

General